FIRST EDITION

A ∫troll through calculus

A guide for the merely curious

Anthony Barcellos

American River College
Sacramento, California

Bassim Hamadeh, CEO and Publisher
Michael Simpson, Vice President of Acquisitions
Jamie Giganti, Managing Editor
Jess Busch, Senior Graphic Designer
Kristina Stolte, Acquisitions Editor
Michelle Piehl, Project Editor
Alexa Lucido, Licensing Coordinator
Layout and design by Butow Communications Group
http://www.butow.net

First published in the United States of America in 2015 by Cognella, Inc.

ISBN: 978-1-63189-801-3 (pbk) / 978-1-63189-802-0 (br) / 978-1-63487-179-2 (sb)

www.cognella.com 800-200-3908

0 Preface

This book is definitely for you. After all, you're browsing the first pages of a book whose title contains the word "calculus," so you're obviously curious. The pages that follow will help to satisfy your curiosity. Most people think calculus is super-difficult math for geniuses. It's not. Calculus is about measuring things and how fast they change. The math is often very simple and readily accessible to curious, intelligent people.

I will be your tour guide during this stroll through calculus—not your lecturer. I have deliberately kept the discussions in this book at an elementary level. In fact, I'm going to start with something as simple as the area of a rectangle. That should be a friendly entry level for almost everyone. It would be nice if you're not too frightened of a little algebra, and later on there will be places where some knowledge of trigonometry might help, but in every case I will lead you through step by step. If I start explaining things you already know, then you can skim those parts until you get to material that is more engaging and less obvious. My main concern is to avoid assuming too much mathematical knowledge on the part of the reader.

Calculus is such an important part of math and science that it's a shame more people aren't acquainted with it. That's what I intend to provide here: an acquaintance. This introduction won't make you into a calculus expert, but it will give you some appreciation for what calculus is, how it's used, and why it's important. You'll run into some famous names as we examine the concepts of calculus because I occasionally try to put things in historical context by giving due credit to the clever people who initially explained or discovered the key ideas. Calculus has ancient roots that go back even as far as Archimedes, the greatest of its early pioneers.

A Stroll through Calculus is the kind of book I would have been delighted to find when I first began to learn about calculus. Since it didn't exist, I finally had to sit down and write it.

How to read this book

The best way to read this book is with a pen or pencil in your hand and a tablet of paper nearby. You'll get the most out of the examples I present if you check the steps yourself. I try not to leave too much out, but the author always has to perform a balancing act between brevity and detail. A short explanation can be sweet, but terse discussions are often difficult to follow, while a detailed explanation can be easy to follow while also being boring and clogged with every possible tiny step. My plan is to always provide enough details and steps so that you can follow the action, although you may occasionally find it useful to pause to verify what I'm saying with a line or two of calculation on your notepad.

Participatory reading is the best way to understand what we encounter on our stroll through calculus.

Dedication

This book is in memory of two men of mathematics:

Clyde M. Wilcoxon (1929–2004), late professor emeritus of mathematics at Porterville College. Every day in the classroom I consciously and unconsciously draw on his lessons. He taught me math and he taught me teaching.

Edmund Silverbrand (1912–2005), glider pilot, raconteur, teacher, principal, education lobbyist, and adult school advocate. In all respects, he was the quintessential gentleman scholar.

Table of Contents

Credits

1 The Area of a Rectangle

Thinking inside the box

The area of a rectangle is length times width. What could be simpler? Calculus can be used to find the area of more complicated regions, so we're going to work our way up to calculus by starting with the simplest case we can. The first surprise is that the area of a rectangle, as simple as it is, can represent a lot of different things—not just area.

Let's take a rectangle whose length is 4 feet and whose width is 3 feet. According to the length-times-width formula, its area is $4 \times 3 = 12$, which has a nice illustration to go with it. Count the squares in Figure 1-1. There are 12 of them.

Figure 1-1.

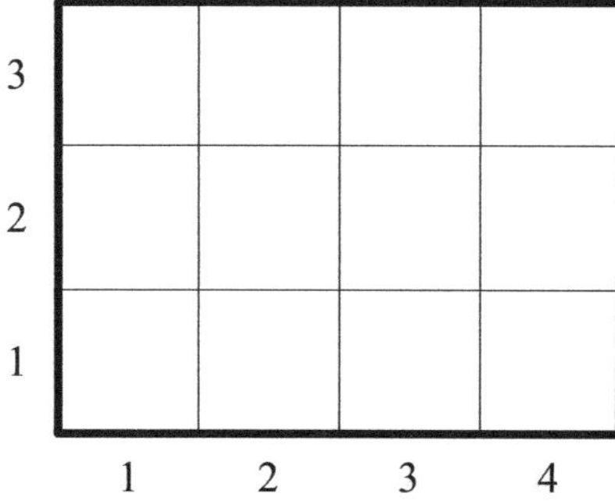

What does each square represent? Since I gave the units of measurement in terms of feet, each square in Figure 1-1 is a square foot. When a square is 1 unit long on each side, we call it a unit square. If we write the area computation again, this time including the measurement units, we have (4 feet)(3 feet) = 12 square feet. As you'll recall from algebra, we usually prefer to indicate a square with a superscript, like this: 12 feet^2. (Here you can see one of the reasons why I'm going to avoid footnotes in this book.) We're therefore using feet to measure length and feet^2 to measure area.

Figure 1-2.

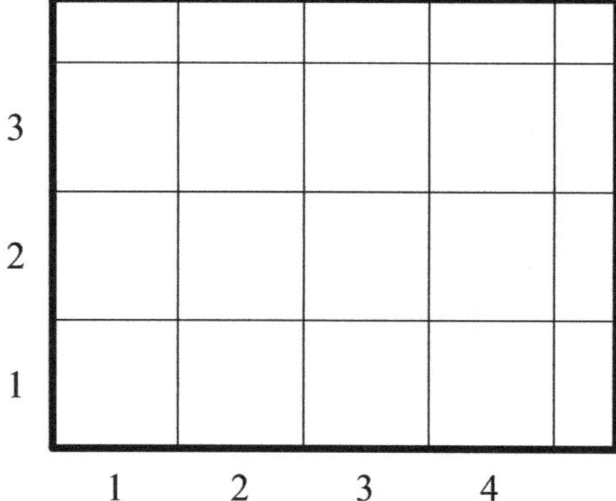

The calculation of area works pretty well with fractions, too. Suppose your rectangle has length 4.5 feet and width 3.5 feet. The area would be (4.5 feet)(3.5 feet) = 15.75 feet^2. Check out Figure 1-2 and try to count the squares. The original 12 are still there, but now we have some half squares (little rectangles) to count, as well as the small square in the upper right-hand corner, which is $\frac{1}{4}$ of a regular square. Each pair of little rectangles equals one of the unit squares, so we can count up our results, as shown in Figure 1-3.

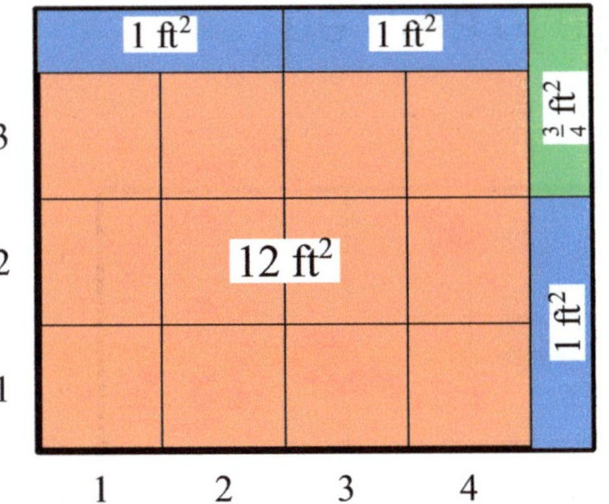

Figure 1-3.

To save some space, we're using the traditional abbreviation ft^2 for $feet^2$. In addition to the original 12 ft^2, we have two square feet across the top, one square foot on the lower right edge, and the remaining $\frac{3}{4}$ ft^2 on the upper right edge. Therefore the total is $12 + 2 + 1 + \frac{3}{4} =$ 15.75 ft^2. Naturally we could have coaxed this result out of a calculator, and some of us may recall the nearly lost art of performing decimal multiplication by hand, but checking the results by counting squares in Figure 1-3 makes the example nice and concrete. The picture matches the calculation.

Thinking outside the box

I began by saying we were going to talk about area, and we have, at least for the rectangle. We could now move on to other areas you might recall—things like triangles, trapezoids, and circles—but we're going to stay with rectangles for a while longer. Before moving on to other calculations of area, I need to make an important point about the calculations we just made. *They don't have to be about area.* While we may compute something that seems to be the area of a geometric object, it could actually represent something very different. Let me show you.

Consider our original rectangle again, but let's change the measurement units. We'll keep the numbers 4 and 3, but we're going to use dimensions other than length. While this may sound strange, the result will be quite familiar. Suppose that the "length" of the rectangle is now given as 4 hours (that's right, I said *hours*) and its "width" is now 3 miles per hour. Using traditional abbreviations for the units of measurement, this is what we get when we compute the "area" again:

$$(4 \text{ hr})(3 \text{ mi/hr}) = 12 \text{ mi}.$$

Notice how the "hr"—the hours—canceled and left us with miles as the units of measurement. If that looks unfamiliar, this may help, remembering that you can rearrange factors in a multiplication:

$$(4\text{ hr})\left(3\,\frac{\text{mi}}{\text{hr}}\right) = (4)(3)\left(\cancel{\text{hr}}\,\frac{\text{mi}}{\cancel{\text{hr}}}\right) = 12\text{ mi}.$$

We did the same computation as for area, but this time we ended up with a distance for our answer. Galileo observed hundreds of years ago that area calculations can be used to find distance, so we are simply following in his footsteps. You probably recognize that our answer seems reasonable enough: If you walk for 4 hours at 3 miles per hour, you'll travel 12 miles. This is a well-known formula: Distance equals Rate times Time, often written as $d = r \times t$.

If we already knew the formula for computing distance from rate and time, why did we bother pretending it's the same as an area computation? Well, there's no pretense about it. Math gets a lot of its power by taking advantage of equivalent quantities. We often find that a problem which seems difficult in a particular context can become quite easy when approached in a different context. As long as we're willing to play with the units and get away from the idea that an area can represent only an area, we begin to see the power of generalization.

The box itself

Let's repeat our "area" calculation one more time, but instead of obtaining an area or a distance, we'll see that it can produce yet another quantity. This time we'll use a "length" of 4 feet (okay, that really is a length) and a "width" of 3 feet^2 (which you can see is already an area). When we multiply our quantities together, we get

$$(4\text{ ft})(3\text{ ft}^2) = 12\text{ ft}^3.$$

What does this mean? We multiplied a length by an area and got a volume (cubic units, as they say). If you're wondering how we could represent this calculation physically, it's simply the volume of a solid object. While it could correspond to a number of different objects, the simplest example is a nice rectangular box, as shown in Figure 1-4. The actual length of the box is 4 feet, while the area of the side is 3 ft^2.

Figure 1-4.

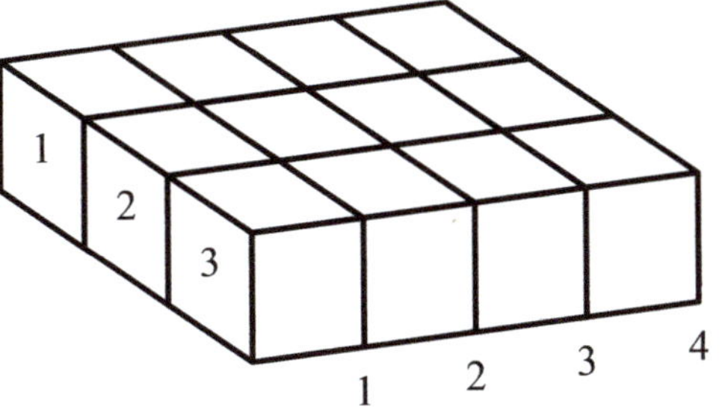

Before we leave our rectangular calculations, even though the results may be anything but conventionally rectangular, let's set up a general representation that will serve us well in the following chapters. I'm talking about the Cartesian coordinate system, the rectangular grid that got its name from René Descartes, the French mathematician who famously liked to spend his mornings lying in bed. He supposedly invented rectangular coordinates and other features of what we now call analytic geometry during moments of wakefulness while sleeping in. A hero to many mathematicians who are jealous of him, Descartes is the perfect example of someone who "should of stood in bed." His discoveries made him too famous for his own good, to the point that Queen Christina of Sweden was inspired to offer him a ridiculously big salary to come to Scandinavia to tutor her. Descartes discovered too late that the queen was an exponent of vigorous outdoor activities and preferred to receive her math lessons while riding horseback in the early morning hours. Perhaps too proud to admit his mistake promptly and return to the less chilly environment of France, Descartes caught pneumonia and died. Let that be a lesson to us all.

The Cartesian coordinate system has long outlived its inventor, providing us with a perpendicular pair of number lines—the coordinate axes—against which we can plot various shapes and figures. In Figure 1-5, we have plotted several different points on the Cartesian grid, where each point is determined by a pair of numbers called its *coordinates*. The first number in each pair is the horizontal or x coordinate, and the second number is the vertical or y coordinate. As you can see in the figure, the point (4, 2) is not the same as the point (2, 4), so the order of the numbers is important. A few of the points are highlighted in red. They all share a vertical coordinate of $y = 3$ and therefore lie on a common horizontal line.

Figure 1-5.

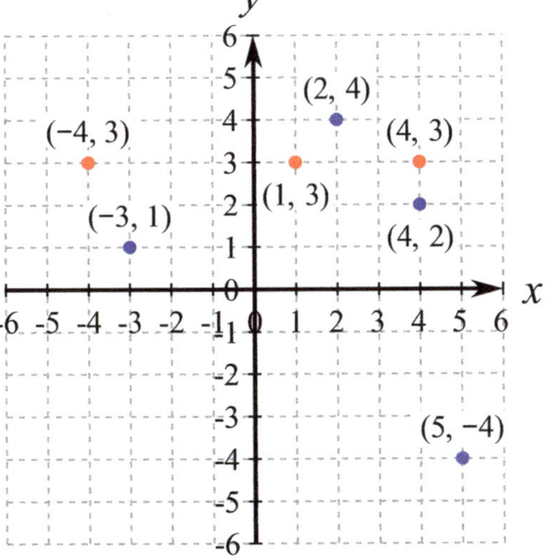

In Figure 1-6, we have graphed the horizontal line $y = 3$ and marked off the region underneath the line for values of x between 0 and 4. Although the letters x and y are the traditional labels for the horizontal and vertical axes, the labels can be changed to whatever is more convenient. For example, in the distance example we did earlier, we used hours for the horizontal measure and miles per hour for the vertical measure. People usually change the label on the horizontal axis to t (for time) in that case, and we'll feel free to do that whenever it's convenient.

Embedding a rectangle in a Cartesian coordinate system is a fairly fancy way to depict it, turning the rectangle into an area under the graph of a line, but we have a good reason for doing it. The rectangular coordinate system is going to be our favorite place for working with calculus, so it helps us to get acquainted with it in a simple situation before we push on to more challenging cases.

Speaking of more challenging cases, now it's time to move on to triangles.

Figure 1-6.

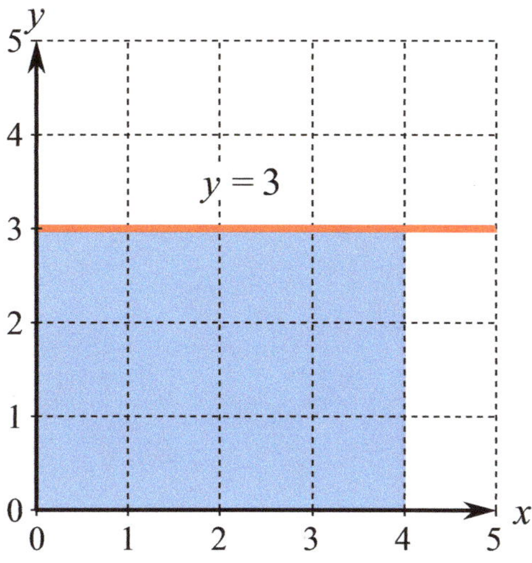

2 A Loved Triangle

Sometimes half is all you need

It's time to add the right triangle to our collection of mathematical tools. I call it a tool because we will be able to use it to solve problems. We can begin by noticing how closely related the right triangle is to the rectangle. In fact, it's half of it.

Suppose we start out by labeling the dimensions of a rectangle as Base and Height instead of Length and Width. We've already seen that it's up to us to decide how to label things and what measurement units to use, so we should feel free to label the rectangle as shown in Figure 2-1. The figure also illustrates how we can break a rectangle along one of its diagonals to create two identical right triangles, where the adjective "right" refers to the fact that the triangle has a right angle (90 degrees) in one of its corners. The original rectangle's area will be Base times Height, of course, so each of its halves has an area equal to $\frac{1}{2}$(Base)(Height). If we use the usual abbreviations of b and h for Base and Height, we can write the formula for the area of a right triangle as

$$\text{Area} = \frac{1}{2}bh.$$

Figure 2-1.

Height

Base

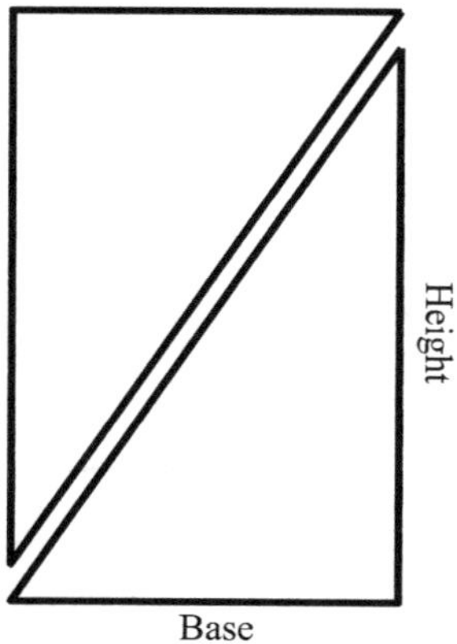

If you seem to recall that this formula applied to triangles in general, and not just right triangles, you're correct. We'll talk about that some more later, but for right now all we need is the right triangle.

Going on a trip

Let's use our area formula for a right triangle to revisit the distance-rate-time calculation that we worked out in the previous chapter. In the

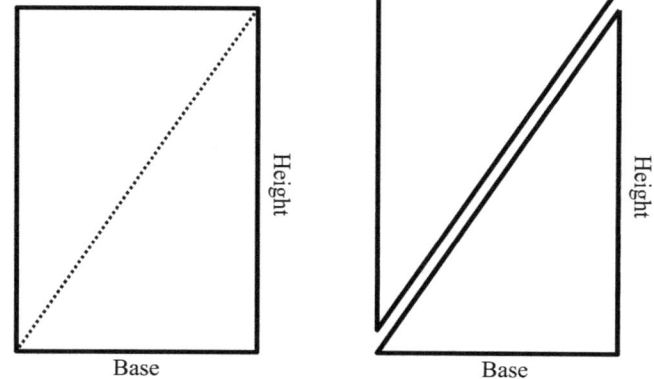

earlier case we used a rectangle as a model for motion at a constant velocity. We can now do something significantly more complicated with essentially the same amount of work. The triangle in Figure 2-2 is the key.

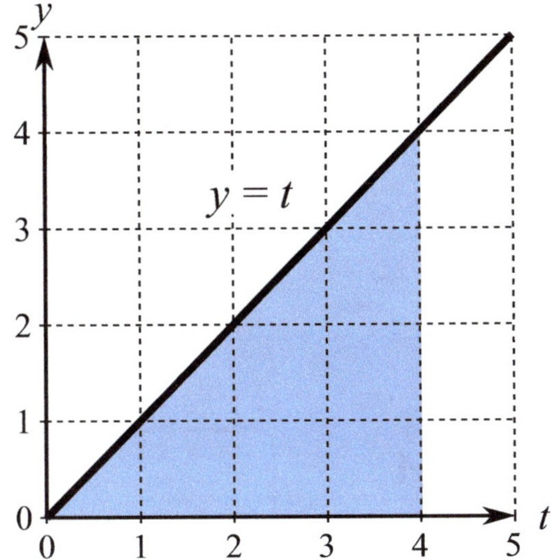

Figure 2-2.

There is a lot of information in Figure 2-2. First of all, notice that we have used t (for time) as the label for the horizontal axis. The diagonal line on the grid can be written as the equation $y = t$ because all of the points on it have equal coordinates. Some examples of points on the line are (0, 0), (1, 1), (2, 2), (3, 3), and (4, 4).

I want to interpret this graph as a record of an object in motion. Let's say that the units of measurement for the horizontal axis are seconds and that the units of measurement for the vertical axis are meters per second. The line $y = t$ merely represents the information that at time $t = 0$ seconds the object was at rest, with velocity $y = 0$ meters per second. Similarly, we can say that the velocity at $t = 1$ sec was $y = 1$ m/sec, at $t = 2$ sec it was $y = 2$ m/sec, and so on. If we wanted, we could include any number of points. How fast was the object traveling at $t = 2.5$ sec? Easy: its velocity was $y = 2.5$ m/sec. And so on and so forth.

When we worked out the corresponding problem with a rectangular representation of constant velocity, we saw that the distance traveled was equal to the number we obtained from an area computation. Let's do that with the triangle in Figure 2-2. The question we are trying to answer is "How far does an object travel in four seconds if its velocity at time t is given by $y = t$?" When we calculate the area of the right triangle depicted in blue in Figure 2-2, we get

$$\text{Distance traveled} = \text{Area} = \frac{1}{2}bh = \frac{1}{2} \cdot (4\text{ sec}) \cdot (4\text{ m/sec}) = 8\text{ m}.$$

This result is quite a bit more advanced than $d = rt$, the formula for distance traveled at a constant velocity. We computed the distance traveled by an object that was constantly speeding up. Let's do it again, this time with a more ambitious case. Instead of using $y = t$, let's try

$y = 2t$. Let's keep the travel time at 4 seconds. If we can figure out how to represent the distance as a triangular area, we can compute it. (If you remember graphing the equations of lines in your algebra class, this should be especially easy for you.) The graph we need is in Figure 2-3.

Figure 2-3.

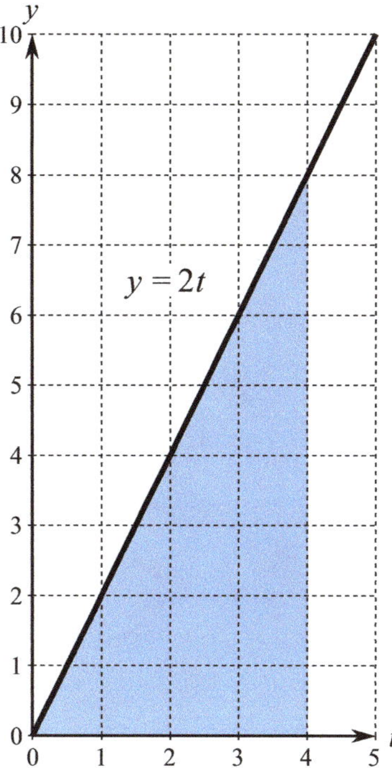

The base of the triangle is 4 seconds, as before, but the height of the triangle is 8 meters per second. When we compute the distance (the area), we get

$$\text{Distance} = \frac{1}{2}bh = \frac{1}{2}\cdot(4\text{ sec})\cdot(8\text{ m/sec}) = 16\text{ m}.$$

It's time to use our tool of triangle area to work out a classic result from physics. This is an important step in working with arbitrary quantities, representing numbers with symbols instead of actual numerical quantities. The result will be a general formula that can be used over the broadest possible range of situations. By taking this step, we're preparing ourselves for some of the computations that we'll encounter later as we get better acquainted with calculus.

Life on the road

To begin, take a look at the velocity equation $y = at$. We've already worked out the cases where $a = 1$ and $a = 2$, but now we're using a as a generic quantity that we could fill in later, if we want. Instead of choosing a particular duration—such as 4 seconds, our choice in the two previous examples—we'll let t represent a generic time, perhaps to be specified later. It's not really that difficult to draw the graph that represents our generic situation. Check out Figure 2-4.

Figure 2-4.

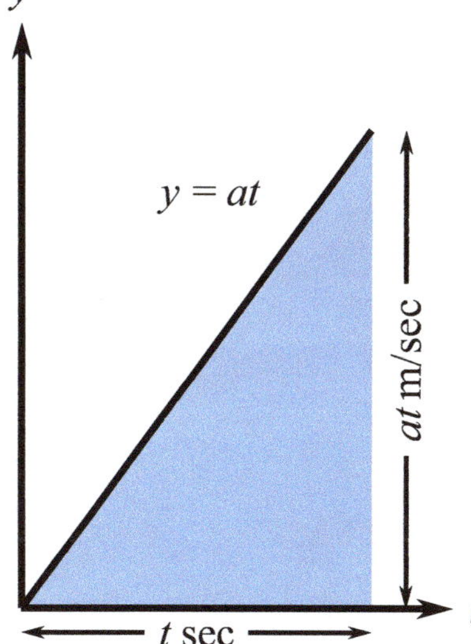

One of the first things you'll notice is that the numbers are missing from the time and velocity axes. Since we haven't specified the exact quantities we're using, we can't really mark them on the graph. We do know, however, that if the base of the triangle is t seconds (that's our generic time), then the corresponding height is $y = at$ meters per second (as given by our velocity equation). Since we know our generic base and height, we can compute our generic result:

$$\text{Distance} = \frac{1}{2}bh = \frac{1}{2}(t)(at) = \frac{1}{2}at^2.$$

As I said, this is an important result from physics, which you may have seen before in a natural science class. Galileo worked it out decades before calculus was invented. It is the formula for distance traveled by an object moving under constant acceleration—that is, an object that is speeding up at a constant rate. This formula includes both of our previous results, as we now confirm. Check what happens when we plug in $a = 1$ with $t = 4$:

$$\text{Distance} = \frac{1}{2}at^2 = \frac{1}{2} \cdot 1 \cdot 4^2 = \frac{1}{2} \cdot 16 = 8.$$

That agrees with the $y = t$ example. If we plug in $a = 2$ and $t = 4$ to check our $y = 2t$ example, we get

$$\text{Distance} = \frac{1}{2}at^2 = \frac{1}{2} \cdot 2 \cdot 4^2 = 16.$$

which agrees with what we got before. We can plug in any numbers we like for other examples of constant acceleration. The formula covers them all. We could, if we wished, change the units, too. Feel free to use the formula with hours and miles per hour, if you wish; just keep the units consistent with each other.

When triangle met rectangle

Before we take a break from the triangle, we are going to address two additional matters. The first is a special geometric shape which I will call the *vertical trapezoid*. The vertical trapezoid is just a rectangle with a right triangle neatly set on top. We will have occasion to use the vertical trapezoid in some future computations, and we now have the perfect tools to compute its area for future reference.

As we see in Figure 2-5, the vertical trapezoid combines a rectangle with base b and height h with a triangle of identical base b and a height $H - h$.

Figure 2-5.

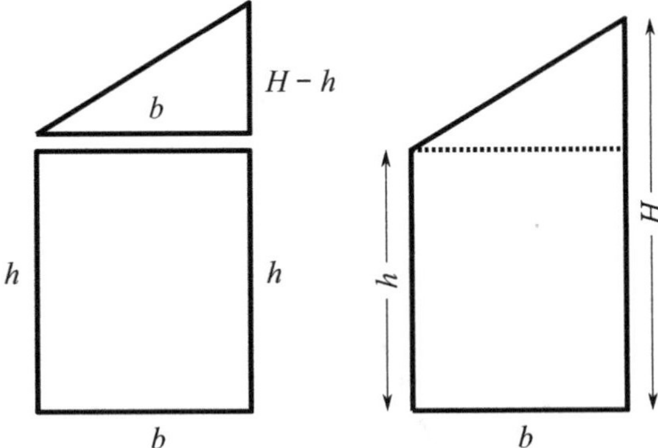

As you can see in Figure 2-5, I chose $H - h$ for the height of the triangle so that the total height is H when we stack it on top of the rectangle. The area of the vertical trapezoid is simply the sum of the rectangle's area and the triangle's area:

$$\begin{aligned}\text{Area} &= bh + \frac{1}{2}b(H - h)\\ &= \frac{2}{2}bh + \frac{1}{2}b(H - h)\\ &= \frac{1}{2}b(2h + H - h)\\ &= \frac{1}{2}b(h + H).\end{aligned}$$

That's the most algebra we've done so far, but it's given us the area of a vertical trapezoid, one more tool for our collection.

Triangles in general

As I mentioned earlier, the area formula we found for right triangles is valid for all triangles. It's not too difficult to see why it works. Check out the family of triangles in Figure 2-6. All of them have the same base and the same height, but their shapes are different. Why do they all have the same area? It's an application of Cavalieri's principle, a discovery made by an Italian mathematician while he was trying to compute the volumes of wine barrels. Cavalieri deduced that any two objects with identical cross-sections would have equal measures. If we look at the triangles again, we see that each one has the same horizontal cross-section for a given height. One example of the common cross-section is shown in Figure 2-7.

Figure 2-6.

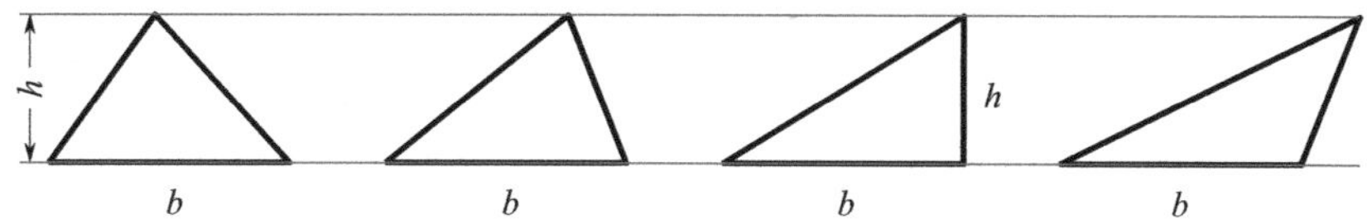

Figure 2-7.

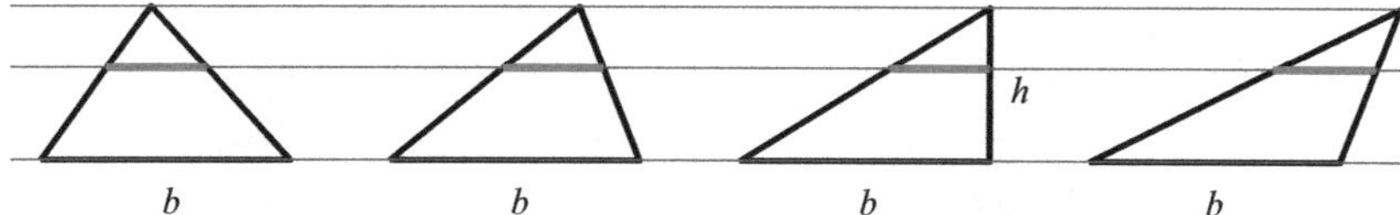

One of the simplest illustrations of Cavalieri's principle is a deck of cards. When stacked up nice and square, a deck of cards has a volume easily computed as the volume of a box with rectangular sides (Volume equals Length times Width times Height). If, however, the cards are skewed rather than neatly stacked, the deck still has the same total volume. Nothing was lost or gained by rearranging the pile.

The same kind of reasoning applies to a stack of coins. As shown in Figure 2-8, it doesn't matter if the stack of coins is straight or crooked. The volume is the same.

Figure 2-8.

While Cavalieri applied his principle to the calculation of volumes, we just used it with triangles to show the equality of areas. This is one of the themes of mathematics, as you'll recall, the notion that powerful ideas have general applicability.

With the formulas we have revisited and rediscovered so far, we can start to write them in traditional calculus notation. That is the topic of the next chapter.

3 Sometimes a Great Notation

How you say it can be as important as what you say

My students prick up their ears when I announce that I'm going to tell them the secret of mathematics. They lean forward a bit, hoping I might be about to let them in on something significant. Then I tell them. *Laziness*.

Naturally, they regard my revelation with suspicion. Some peer at me with narrowed eyes, aware that I must have some point that I will be trying to make. Others grin because that's what you do when it appears the teacher has said something that might be a joke. A few remember Descartes and his habit of lying in bed all morning; if it's a 7:00 a.m. algebra class, they may even exhale a sigh of jealousy.

My usual follow-up is a question: "Which looks simpler?" I then write two things up on the board:

Four plus Five equals Nine

$$4 + 5 = 9$$

They unanimously but tentatively agree that the second statement is simpler. Usually by now everyone in the class is highly suspicious, but my point isn't that esoteric. I make it explicit for them:

"Mathematicians like to keep things simple whenever they can. They use symbology to make their lives easier."

Which would you rather do, multiply 9 by 12 to get 108 or multiply IX by XII to get CVIII? How would you even do the latter? Can you do multiplication in Roman numerals? Clearly our adoption of Arabic numerals in place of the Roman version must stand as one of the great milestones in mathematical progress. Let me take this just a little bit further.

What does the numeral 2 mean to you? We can all agree that it represents the number two and is convenient for counting twins or shoes or other things that come in pairs. But what does the 2 in 125 represent? Because it's in the column reserved for tens, it represents 20, not just 2. The position of a numeral in our positional notation makes all the difference in the world. It's a very powerful yet simple way to represent numbers of all magnitudes, large or small, with only ten digits and a decimal point. The Romans had to work much harder at their math than we do today, so thank the lazy mathematicians who came up with the improvement.

The best of all possible notations

Thanks to Gottfried Wilhelm Leibniz, calculus notation is one of the great labor-saving devices in mathematics. Leibniz worked diligently to make his notation as logical and natural as possible. As a direct consequence of his work, the European mathematicians rapidly outpaced their English counterparts. The English felt obligated to stick with the elaborate and difficult style of their countryman Newton, while the Europeans took advantage of Leibniz's elegant and intuitive notation. We're going to look at some of that notation now. Our continuing goal is to understand what it represents and what we can do with it.

Figure 3-1.

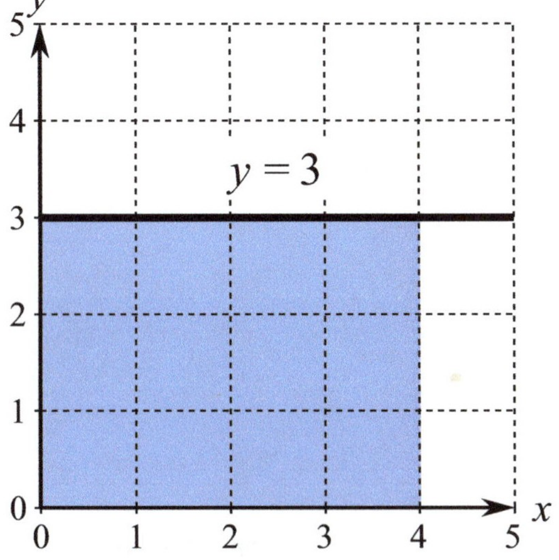

Check out Figure 3-1, which should seem familiar to you. In this graph, we have $y = 3$, and x varies from 0 to 4. Leibniz created a symbol resembling the letter S to indicate the shaded area below $y = 3$ for x between 0 and 4. We already know that the result is supposed to be 12, so I'll write the whole thing out for you:

$$\int_0^4 3 = 12.$$

Note how it's written. The tall symbol is called the *integral symbol* or the *integral sign*, and the lower and upper values of x are recorded as subscripts and superscripts, respectively, on the symbol. (The lower and upper values of x are called the *limits of integration*.) Immediately after the symbol we write the expression or formula (called the *integrand*) that tells us the height of the area we're finding. Let's try a couple of triangular cases.

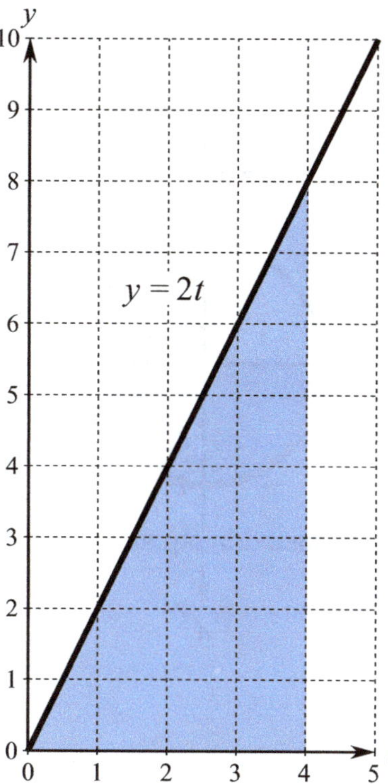

Figure 3-2.

The graph in Figure 3-2 consists of $y = 2t$ for the line bounding the top of the triangular region. We're using t in this example, where the bounds on t are 0 and 4: $0 \le t \le 4$. Recall that our result for the triangular area was 16, so we can use the integral symbol to write a calculus equation:

$$\int_0^4 2t = 16.$$

If we go back to the example with the generic formula and results, the one that gave us the equation from physics for distance traveled during constant acceleration, we put in $y = at$ for the right-hand part of the integral symbol (the *integrand*), and we use the lower and upper values of 0 and t for the sub- and superscripts (the *limits of integration*). The result is the calculus equation

$$\int_0^t at = \frac{1}{2}at^2.$$

See Figure 3-3 for a reminder of the graph for distance traveled under constant acceleration.

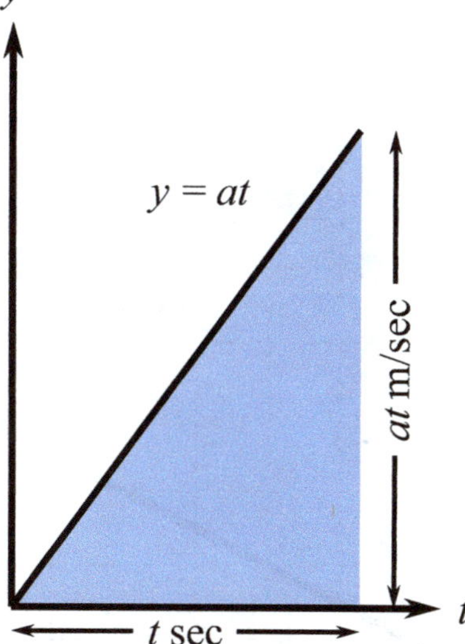

Figure 3-3.

Breaking new ground

All we've done so far is recycle examples from earlier chapters to illustrate the use of our new notation. We can also use Leibniz notation to record some new calculations.

We begin by graphing the line $y = \frac{1}{2}x + 4$ on a Cartesian coordinate system and shade the area below the line for x values between 2 and 6.

If we take a look at the graph in Figure 3-4, we can see that the shaded region is in the form of a vertical trapezoid. The base is simply $b = 4$, where $2 \leq x \leq 6$, while the short height is $h = 5$ and the tall height is $H = 7$. That's enough information to allow us to find the area:

$$\frac{1}{2}b(h+H) = \frac{1}{2} \cdot 4(5+7) = 24.$$

If we put these pieces together, we have $\frac{1}{2}x + 4$ for the integrand and $x = 2$ and $x = 6$ for our limits of integration (the sub- and superscript on the integral sign). Our Leibniz notation equation is

$$\int_2^6 \left(\frac{1}{2}x + 4\right) = 24.$$

The parentheses are there to group the terms of the integrand because there are more than one of them. (We want people to know the 4 is part of the integrand and not just a number being added to the integral.)

Figure 3-4.

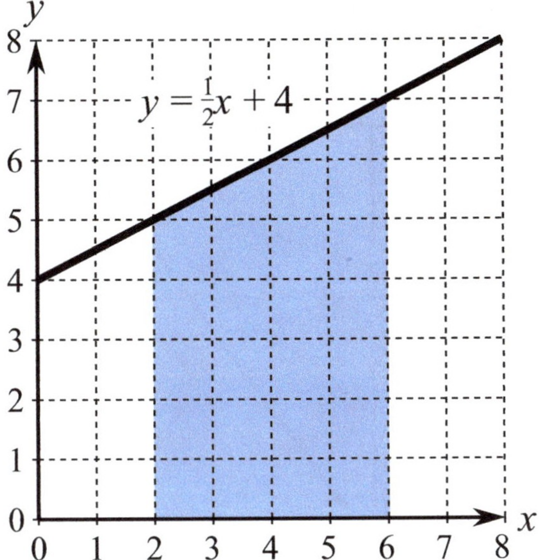

If you remember your algebra pretty well, we can look at one more situation, just to get away from all of these examples with straight-edge boundaries. We learned in algebra that the equation $x^2 + y^2 = r^2$ corresponds to a circle of radius r (with its center at the origin (0, 0)) if you graph it on a Cartesian coordinate system. Thus $x^2 + y^2 = 4$ would give us a circle of radius 2. By solving for y^2, we get $y^2 = 4 - x^2$, so it must be that $y = \pm\sqrt{4 - x^2}$. If we choose the plus sign, so that y is not negative, we get the top half of the circle of radius 2; that is, $y = \sqrt{4 - x^2}$ has a semicircle of radius 2 as its graph, as shown in Figure 3-5. We know that the area of a circle is πr^2, so the semicircle's area is $\frac{1}{2}\pi \cdot 2^2 = 2\pi$. We can write a very interesting-looking Leibniz equation:

$$\int_{-2}^{2} \sqrt{4 - x^2} = 2\pi.$$

Figure 3-5.

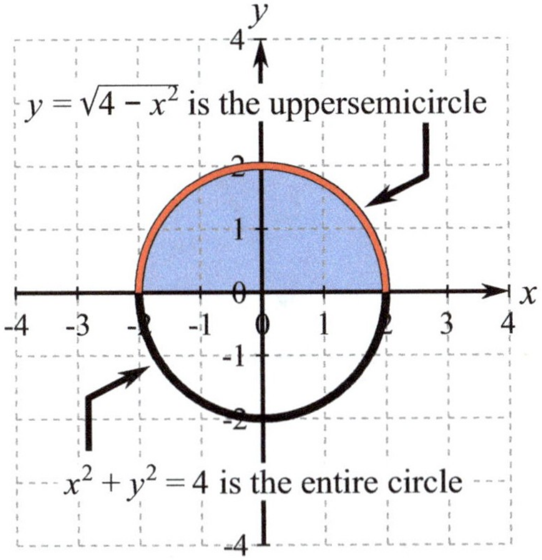

This is all very good, but you probably can tell that calculus isn't done this way. We don't usually find the value of an integral by drawing a picture and computing an area. If the integral is going to be a useful tool for finding areas (and distances and volumes and other things that can be represented as areas), we need better techniques for evaluating them. We will start to acquire those techniques in the next chapter, which begins with an exercise in pattern recognition.

Postscript

If you've seen the integral sign before, you may be thinking that I left something out of the traditional integration notation. You're right. I'm using a simplified version because that's all we need for now. I'm saving the more detailed symbol for later, when it will serve a purpose.

4 Simplicity Patterns

Can you guess the next number?

Although Newton and Leibniz get credit today for independently creating calculus, we should not think that calculus happened overnight. Both men were working with the legacies of their many mathematical predecessors. Newton himself, in a moment of uncharacteristic modesty, wrote, "If I have seen a little further it is by standing on the shoulders of Giants." Calculus as we know it today is a brilliant synthesis of many disparate efforts by many different researchers, melded wonderfully together by the keen insight of geniuses.

In this chapter, we are going to gather together some of the preliminary results that caught the eyes of the calculus pioneers. They saw patterns in the results. Sometimes it was possible to create one general rule that covered many different individual examples. Let's look at some of those examples now, and I will show you one of the patterns that began to reveal the outlines of the calculus that was to come. On the way, we will meet one of history's greatest mathematical giants, Archimedes himself.

When we talked about areas before, I often used 4 (with some suitable units) as one of the dimensions. That was just for convenience of illustration, because 4 is just big enough to be interesting. Now, however, I'm going to admit that using 4 obscured a pattern that will pop up before our eyes in the simpler cases I'm about to show you. In these next examples, our base will be of length 1. Check out the graphs in Figure 4-1. In each case I'm using $0 \leq x \leq 1$ on the horizontal axis. The equations in the graphs are $y = 1$, $y = x$, $y = x^2$, and $y = x^3$. If you remember that $1 = x^0$, then you can see that my choice of equations is just a sequence of increasing powers of x. What is the integral for each of these expressions for x between 0 and 1?

Figure 4-1.

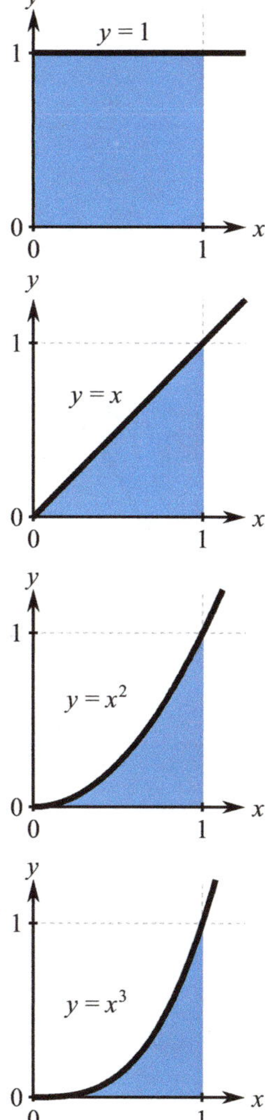

Well, we already know some of them, right? For the first graph, the area is just a unit square, so the integral is 1. The second graph has a right triangle whose area is $\frac{1}{2}$. The third graph contains a bit of a parabola, which we have not yet encountered in our stroll through calculus. Fortunately, however, Archimedes got there well ahead of us (he actually beat us by over 2200 years) and figured out just about everything we could want to know about parabolas. In particular, Archimedes determined a parabola's area.

Remember second-degree polynomials from algebra? They're the expressions whose graphs are parabolas, things like $y = x^2 + 5x + 6$, for example. The graph in Figure 4-1 that corresponds to $y = x^2$ is the simplest example of a parabola. We have shaded in the region below the parabolic curve. Suppose we also put a nice rectangular box around it, as in the first graph in Figure 4-2. Without the tools of calculus or even a good system of number notation, Archimedes used his insight into the geometry of the parabola to prove that its curve divides the box into regions where one has two-thirds of the area and the other has one-third of the area. Our parabola example from Figure 4-1 fits nicely inside a unit square, but Archimedes allowed the parabola and its enclosing box to have any proportions (see the other graphs in Figure 4-2). Whatever the dimensions, $\frac{2}{3}$ of the area is above the curve and $\frac{1}{3}$ is below. (I am not coming anywhere close to doing justice to what Archimedes discovered. He found results for parabolas and triangles as well as parabolas and parallelograms. We're just using a special case that serves our present purpose.)

Figure 4-2.

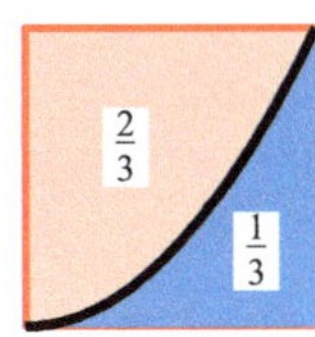

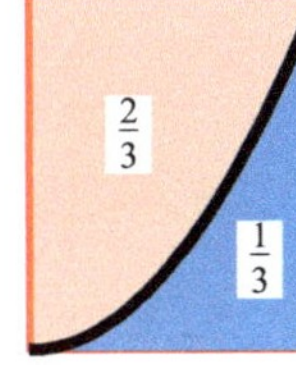

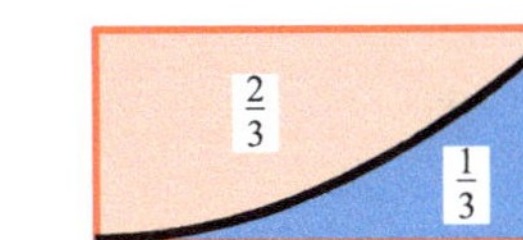

Power up

With the information we've collected so far, we can write the integrals for the first three areas in Figure 4-1:

$$\int_0^1 1 = 1 \qquad \int_0^1 x = \frac{1}{2} \qquad \int_0^1 x^2 = \frac{1}{3}.$$

Would you care to make a conjecture about the area in the fourth graph? In order to continue the pattern we're observing in the first three cases, the fourth integral must be

$$\int_0^1 x^3 = \frac{1}{4}.$$

We might then be brave enough to label our graphs as shown in Figure 4-3.

As a matter of fact, the fourth integral does actually equal $\frac{1}{4}$. They are all special cases of what mathematicians call a *power rule*, in this case the power rule for integrals or, more commonly, the power rule for integration. We can write the power rule for integration in a single formula that covers all the cases we've seen so far and continues the patterns for all the other values we haven't even looked at yet:

$$\int_0^1 x^n = \frac{1}{n+1}.$$

We have actually justified this formula ourselves for only $n = 0$ and $n = 1$. We called on Archimedes for help with $n = 2$. At this point, anyway, you'll just have to take my word for it that it works for other values of n (and n doesn't even need to be a whole number, but some things we need to save for later). We are, however, going to do some further calculations that will help us see that the power rule for integration is reasonable. By the time we're done, I don't think there'll be much doubt about its validity. Next, though, let's try to generalize the limits of integration. Remember that this term refers to the sub- and superscripts on the integral sign. The power rule pattern showed up when we restricted ourselves to x values between 0 and 1, but surely there must be an even more general version of the power rule for other limits of integration. Once we find that more general rule, we will be ready to use our calculus tool box on more interesting problems from geometry and physics.

Figure 4-3.

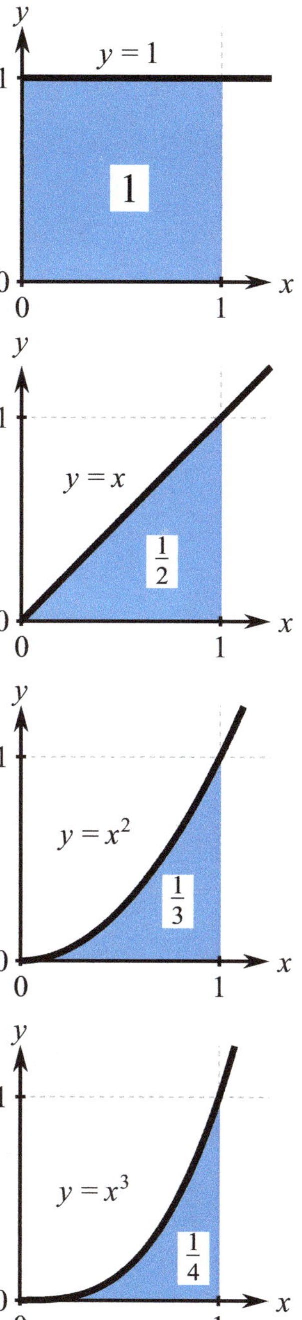

5 Integration from *a* to *b*

We move to a new neighborhood

We launched into some speculation in the previous chapter and came up with a conjecture I called the power rule for integrals. Our results were based on a combination of calculation, help from ancient authority (Archimedes), and guess-work. In this chapter, I'm going to push the speculation further. Then we'll take some time in the next chapter to put our speculation on a firmer footing with some numerical calculations.

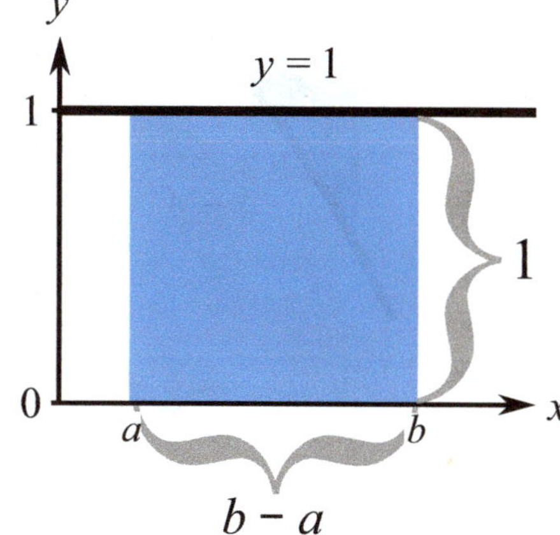

Figure 5-1.

To begin our new pattern building, let's generalize the cases $y = 1$ and $y = x$. Instead of finding the value of integrals for these expressions for x between 0 and 1, let's see what we can do if x is allowed to range between the unspecified values of a and b. To make things simpler, I'll assume that a and b are both positive numbers and that a is smaller than b. We can then compute some integrals with reliable results because our first two cases rely on our old friends the rectangle and the vertical trapezoid. As usual, a nice graph sets the stage for our computation.

As you can see in Figure 5-1, we need to find the shaded rectangular area that lies beneath the line $y = 1$ for x values between a and b. The length of the base of the rectangle is just $b - a$ and the height of the rectangle is just 1, so its area is $(b - a)(1) = b - a$. Our calculus equation is

$$\int_a^b 1 = b - a.$$

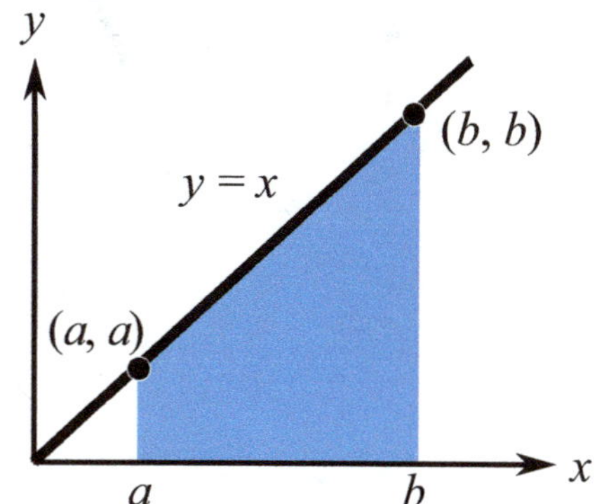

Figure 5-2.

Now we repeat the process, this time with $y = x$ and $a \leq x \leq b$. If you look at the graph provided in Figure 5-2, you'll see that our shaded area under the line $y = x$ for x between a and b is a vertical trapezoid. As with the rectangle in the previous example, the length of its base is

$b - a$. The short height h is a and the tall height H is b. Plugging these into the area formula for a vertical trapezoid, we get an area equal to $\frac{1}{2} \cdot \text{Base} \cdot (h + H) = \frac{1}{2} \cdot (b - a) \cdot (a + b)$. From basic algebra, we know that we can multiply out $(b - a)(a + b)$ to get $b^2 - a^2$. Putting this together with the $\frac{1}{2}$ from the previous expression, our calculus equation is

$$\int_a^b x = \frac{1}{2}(b^2 - a^2).$$

That takes care of the easy stuff. Let's call on our old buddy Archimedes for some assistance with $y = x^2$ and $a \le x \le b$. Now things are getting interesting. First, consider the case where x goes from 0 to b, as shown in the first graph in Figure 5-3. The rectangular box that we inscribed around the parabolic region has a base of length b and a height of b^2. The rectangle's area is therefore $(b)(b^2) = b^3$.

Figure 5-3.

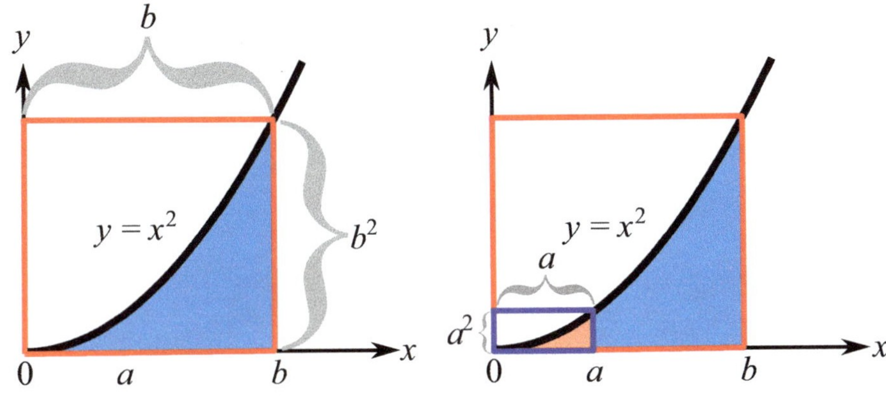

Thanks to Archimedes, we know that the shaded area under the parabola is one-third of the area of the rectangle, so it's $\frac{1}{3}b^3$. But this parabolic area is more than we want. We need to subtract the part between 0 and a. That's the part shown in the second graph in Figure 5-3. It's a box of area a^3 that encloses the portion of the shaded region we need to remove in order to get the parabolic area we actually want.

As before, Archimedes lets us know that the piece to be taken away has area $\frac{1}{3}a^3$. Once we subtract that from $\frac{1}{3}b^3$, we are done. Our result is a new calculus equation that gives us the shaded region in Figure 5-4:

$$\int_a^b x^2 = \frac{1}{3}b^3 - \frac{1}{3}a^3 = \frac{1}{3}(b^3 - a^3).$$

Figure 5-4.

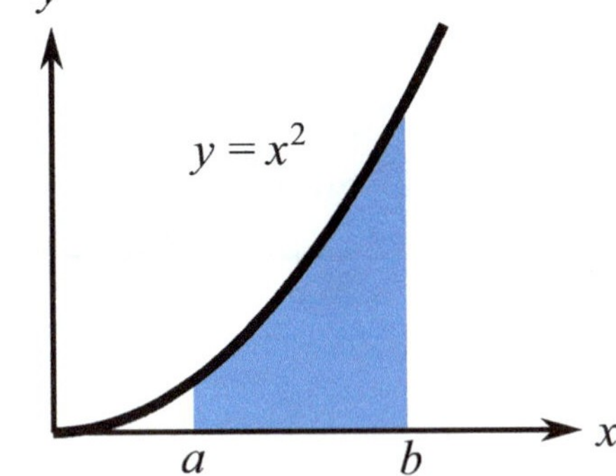

I think we can all agree that a new pattern has been established. It has a lot in common with what we discovered in the previous chapter, but now we can compute integrals for $a \le x \le b$ instead of just for $0 \le x \le 1$. If the pattern continues, then our new and improved power rule for integrals must be

$$\int_a^b x^n = \frac{1}{n+1}(b^{n+1} - a^{n+1}).$$

As we would expect, plugging in $a = 0$ and $b = 1$ reduces the new power rule to the much simpler old power rule. (You may want to verify this.)

6 Numerical Interlude

Archimedes won't mind if we check his work

Do you trust Archimedes? Then perhaps you'd like to skip this chapter and jump ahead to the good stuff in the next one. I intend to pretend that I *don't* trust Archimedes and insist on doing some of the dirty work myself. I invite you to join me in this exercise in verification.

Still here? Then get your calculator out. We're going to crunch a few numbers. While we could accept what other people tell us at face value, it's usually a good idea to examine the evidence ourselves. There are few more distinguished authorities in mathematics history than Archimedes, who told us that $\int_0^1 x^2 = \frac{1}{3}$ (although he himself never saw anything that looked like that calculus equation), but we have the means to verify that statement numerically.

When a mathematician says "verify numerically," that usually means performing computations with actual numbers (quantities) and arriving at an approximate answer. While we might normally prefer an exact answer as opposed to an approximate one, there are cases where an exact answer is simply out of reach. (Not just out of your reach, or my reach, but out of the reach of any mathematician. Fortunately, this is not a book about problems that can't be solved.) Of course, the moment anyone mentions "computations," people are inclined to reach for their calculators. Let's do that now and perform some computations that will help us verify the value that Archimedes gave us for the integral of x^2 for $0 \le x \le 1$.

There is, however, one more caution. What calculator are you using? Many modern calculators will happily do the work of computing an integral for us. If, for example, you have a TI-89 or similar model from Texas Instruments, you can enter ∫(x^2, x, 0, 1) from the keyboard and obtain the result 1/3. The calculator knows the power rule for integrals! And it will tell you the answer (provided you know how to ask the question). Advanced scientific calculators from Texas Instruments, Hewlett-Packard, and Casio all have this feature, but we are not going to rely on it. If we're not going to take the word of Archimedes and leave it at that, we're not going to repose more trust in an electronic device. We're going to use our calculators all right, but we're going to use them to work out results that could actually be done by hand computation (if, admittedly, not as quickly). Later I'll have some further comments about the *deus ex machina* with the liquid-crystal eyes, but for now our focus is...*arithmetic*.

With nothing more complicated than addition, subtraction, multiplication, and division, we're going to use the graph in Figure 6-1 to estimate the value of $\int_0^1 x^2$. If our result doesn't come out close to $\frac{1}{3}$, we're going to be very disappointed.

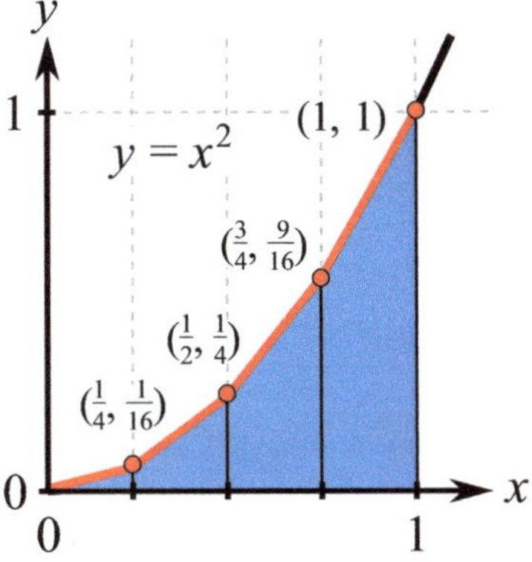

Figure 6-1.

As you can see from the figure, I have replaced the parabolic curve $y = x^2$ with a broken line in four segments. The endpoints of each line segment lie on the parabola. I've also dropped a vertical line from each point down to the x axis. The result is to carve up the region below this parabolic approximation into four pieces: one triangle and three vertical trapezoids, each of which has $\frac{1}{4}$ for its horizontal base. We know how to find their areas and add them up. I've arranged the computations in the accompanying table (see the next page).

Region	Base b	Height h	Height H	Area
Triangle	1/4	1/16	n/a	$\frac{1}{2}bh = \frac{1}{2} \cdot \frac{1}{16} \cdot \frac{1}{4} = \frac{1}{128}$
Trapezoid 1	1/4	1/16	1/4	$\frac{1}{2}b(h+H) = \frac{1}{2} \cdot \frac{1}{4}\left(\frac{1}{16} + \frac{1}{4}\right) = \frac{5}{128}$
Trapezoid 2	1/4	1/4	9/16	$\frac{1}{2}b(h+H) = \frac{1}{2} \cdot \frac{1}{4}\left(\frac{1}{4} + \frac{9}{16}\right) = \frac{13}{128}$
Trapezoid 3	1/4	9/16	1	$\frac{1}{2}b(h+H) = \frac{1}{2} \cdot \frac{1}{4}\left(\frac{9}{16} + 1\right) = \frac{25}{128}$

When we add up the four individual results, we get

$$\frac{1}{128} + \frac{5}{128} + \frac{13}{128} + \frac{25}{128} = \frac{44}{128} = \frac{11}{32}.$$

If we turn $\frac{11}{32}$ into a decimal number, we have 0.34275. That's a pretty good result. Despite our use of very simple calculations and a relatively crude broken-line representation of a nice smooth parabola, we came out with something fairly close to 0.33333.... Do you know why our result turned out to be larger than the true value of the integral we were checking? If you look back at Figure 6-1 again, I don't think you will find the reason too difficult to discern.

Archimedes obtained some initial results similar to ours, but he didn't stop there. He wanted an exact result, not an approximation. Let's follow him partway down his path. Consider how we could get a *better* approximation. How would you do that?

Making it better

I'm sure it didn't take you too long to figure out what to do: use narrower trapezoids, and more of them. I've done the math for you, but it's easy enough to check if you want. The results are summarized in the table below. I began by splitting the interval from 0 to 1 into eight equal pieces so that each trapezoid (okay, the first one is really a triangle, but it's not a problem) has a horizontal base of $\frac{1}{8}$. The points we would be using to draw the broken-line approximation of the parabola would be (0, 0), $\left(\frac{1}{8}, \frac{1}{64}\right)$, $\left(\frac{1}{4}, \frac{1}{16}\right)$, $\left(\frac{3}{8}, \frac{9}{64}\right)$, $\left(\frac{1}{2}, \frac{1}{4}\right)$, $\left(\frac{5}{8}, \frac{25}{64}\right)$, $\left(\frac{3}{4}, \frac{9}{16}\right)$, $\left(\frac{7}{8}, \frac{49}{64}\right)$, and (1, 1). (See Figure 6-2.) I need the y coordinates to provide the heights of the trapezoids for my computations, just as we did before. By the way, although it's usually considered important in mathematics to write things in simplest terms, it's not always an advantage. I filled in the table with all of the heights expressed in terms of 64ths, because a common denominator makes it easier to add up the results.

Figure 6-2.

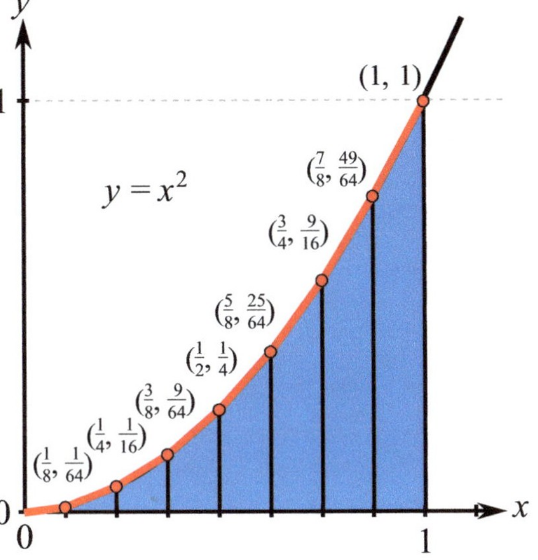

Region	**Base *b***	**Height *h***	**Height *H***	**Area**
Triangle	$\frac{1}{8}$	$\frac{1}{64}$	n/a	$\frac{1}{1024}$
Trapezoid 1	$\frac{1}{8}$	$\frac{1}{64}$	$\frac{4}{64}$	$\frac{5}{1024}$
Trapezoid 2	$\frac{1}{8}$	$\frac{4}{64}$	$\frac{9}{64}$	$\frac{13}{1024}$
Trapezoid 3	$\frac{1}{8}$	$\frac{9}{64}$	$\frac{16}{64}$	$\frac{25}{1024}$
Trapezoid 4	$\frac{1}{8}$	$\frac{16}{64}$	$\frac{25}{64}$	$\frac{41}{1024}$
Trapezoid 5	$\frac{1}{8}$	$\frac{25}{64}$	$\frac{36}{64}$	$\frac{61}{1024}$
Trapezoid 6	$\frac{1}{8}$	$\frac{36}{64}$	$\frac{49}{64}$	$\frac{85}{1024}$
Trapezoid 7	$\frac{1}{8}$	$\frac{49}{64}$	$\frac{64}{64}$	$\frac{113}{1024}$

When we add them up, the total is $\frac{344}{1024} \approx 0.33594$. This is definitely a better result, but not amazingly better than the one before, where we got

0.34275. You'll notice that the digit in the hundredths place has become a 3. If we repeated the process all over again with sixteen trapezoidal subregions, we would pick up another 3 with the answer 0.33398. I've done the calculations up to the case of 64 pieces and have summarized the results in the table below.

n	Result
4	0.34375
8	0.33594
16	0.33398
32	0.33350
64	0.33337

Slowly, but steadily, the approximations are approaching the value $\frac{1}{3}$, whose decimal fraction representation has 3's going on forever. Archimedes was clever enough to find the pattern in the approximations so that he could push on to the exact answer. While this is a common enough practice in calculus, as we will see when we start talking about the slope problem in later chapters, it was an amazing accomplishment in his day.

7 Putting Pieces Together

In any puzzle you start with the straight-edge pieces

If you stayed with me through the previous chapter, you saw more examples of how integration can be considered a process of putting pieces together. We've been doing this with a jigsaw puzzle set of squares, rectangles, trapezoids, and parabolas. Normally I have been careful to do only one specific thing at a time, but now I admit that there was one instance where I played a small trick on you. Do you remember this example?

$$\int_2^6 \left(\frac{1}{2}x + 4\right) = 24$$

Figure 7-1.

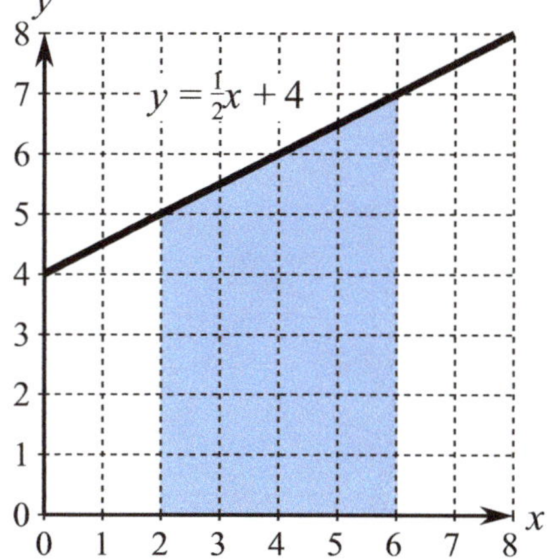

We computed it in an earlier chapter by finding the area of the vertical trapezoid in Figure 7-1. It was also, however, our first example of a two-term integrand, which I neglected to mention at the time. The theme of this chapter is putting pieces together, but I'm going to begin by pulling the pieces of this integral apart, examining them separately, and then putting them back together. When we're done, we will have discovered some very important properties of the integral—properties that will dramatically expand the range and power of our integration tool.

Let's begin by looking at $\int_2^6 \frac{1}{2}x$ and $\int_2^6 4$. These are both easy to evaluate with techniques we've used before. Figure 7-2 contains the graphs we need. As we can see, the region whose area we need to calculate for the first integral is a vertical trapezoid with base $b = 4$, short height $h = 1$, and tall height $H = 3$. Using the area formula for vertical trapezoids, we obtain

$$\int_2^6 \frac{1}{2}x = \frac{1}{2}b(h + H) = \frac{1}{2} \cdot 4 \cdot (1 + 3) = 8.$$

Figure 7-2.

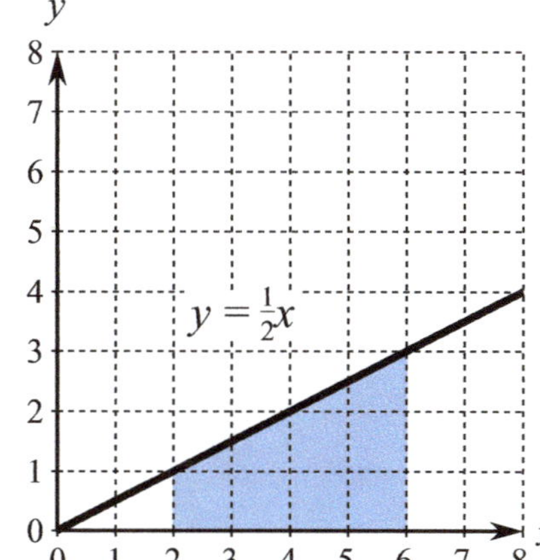

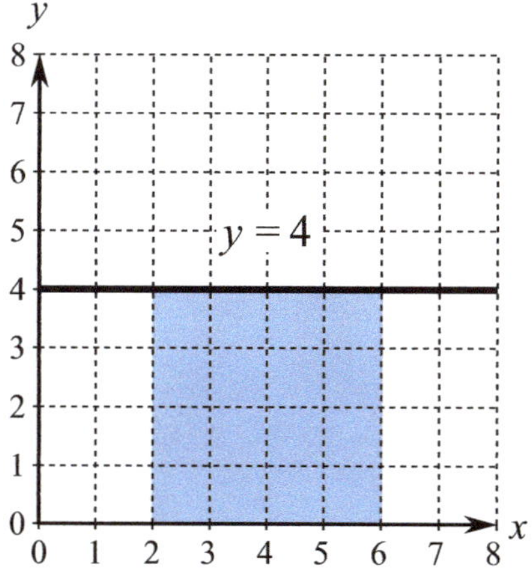

The area we need to calculate for the second integral is just a simple rectangle of base 4 and height 4, so our result is

$$\int_2^6 4 = 4 \cdot 4 = 16.$$

Since 24 = 8 + 16, we have just shown that

$$\int_2^6 \left(\frac{1}{2}x + 4\right) = \int_2^6 \frac{1}{2}x + \int_2^4 4.$$

Do you think that was a coincidence? No, not at all. If an integral has more than one term in its integrand, we can compute integrals for each part separately and add them together. We call this the *addition property of integrals*. In Figure 7-3, I have taken the graphs from Figure 7-2 and illustrated how these results make excellent sense in graphical terms. The area corresponding to one term is simply piled atop the area for the other term.

Figure 7-3.

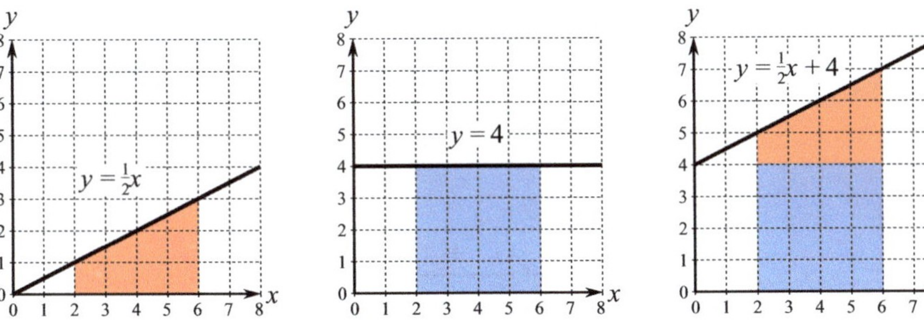

This is a general principle which we need to express with generic symbols. Instead of using specific mathematical expressions for our integrands, let's borrow notation from algebra. In algebra we learned to talk about generic mathematical expressions like $f(x)$ and $g(x)$. (We called it "function notation"—or at least your teacher did—but here it's simply a shorthand for "unspecified math expression.") The general principle we just discovered—the addition property of integrals—can now be written in calculus notation as

$$\int_a^b [f(x) + g(x)] = \int_a^b f(x) + \int_a^b g(x).$$

In ordinary words, this calculus equation merely says, "Hey, you want to compute an integral with two things added together in the integrand? Go ahead and compute the two pieces separately and add the results together." I will now pull a random example out of a hat to try to drive the point home. How about the integral $\int_1^5 (x^3 + x)$? Here we go:

$$\int_1^5 (x^3 + x) = \int_1^5 x^3 + \int_1^5 x = \frac{1}{4}(5^4 - 1^4) + \frac{1}{2}(5^2 - 1^2)$$
$$= \frac{1}{4}(624) + \frac{1}{2}(24) = 156 + 12 = 168.$$

Please check this result for yourself. See how we used the power rule for integrals? Aren't you glad we didn't try to do this by counting squares in a graph or adding up a ton of trapezoidal areas?

Tipping the scale

We're nearly where I want us to be in terms of the properties of the integral, but there is an important factor left. And when I say "factor," I mean it, too. Remember that "factors" is a math term for things that get multiplied together. We just looked at adding stuff up, but now I want to look at what happens when you multiply by a constant factor. For the sake of simplicity, let's start off with a factor of 2. We know that multiplying by 2 is a way to double something. If we multiply an integrand by 2, do we also multiply the value of the whole integral by 2? We most definitely do.

Figure 7-4.

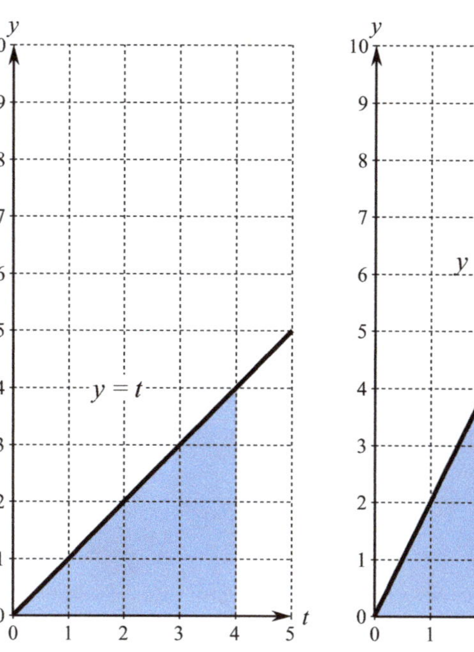

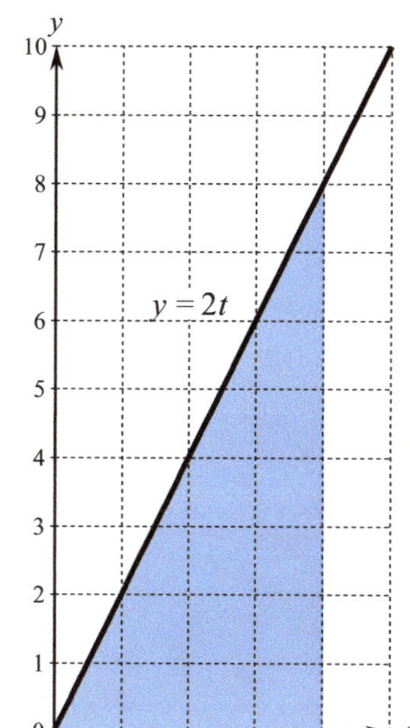

Do you remember the graphs in Figure 7-4? We first saw them in Chapter 2, where the first graph contained a shaded region of area 8 and the second contained a shaded region whose area turned out to be 16. When we look at them now, we can see they represent an important fact, which we can write in calculus notation:

$$\int_0^4 2t = 2\int_0^4 t = 2 \cdot \frac{1}{2}(4^2 - 0^2) = 16.$$

That is, *we can factor out the 2*. Then we can do the simplified integral by applying the power rule for integrals. Neat. If you'd like to see a graphical representation of the doubling, it's not hard to show. Check out Figure 7-5.

Figure 7-5.

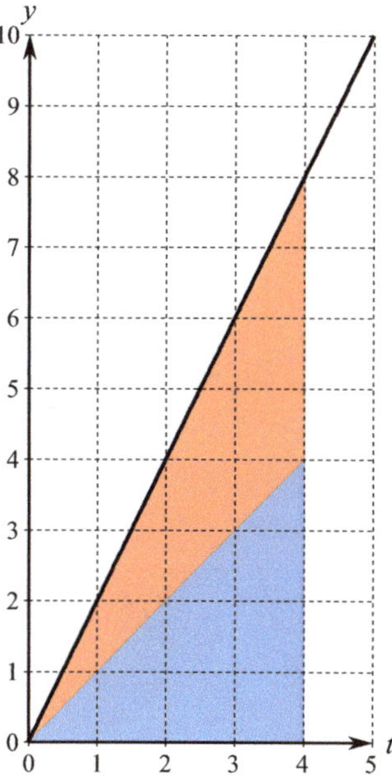

You can see in Figure 7-5 that doubling an expression amounts to piling a copy of the original region on top of itself. The two shaded regions in the graph are of equal area. (Remember Cavalieri? He'd point out that both triangles have identical cross-sections.) The factor of 2 really does cause everything to double—not just the integrand, but the value of the actual integral.

I used 2 because it's easy to visualize, but there's no real reason to limit ourselves to that. We could just as well have chosen 3. Or 4. Or 3.14. In a way, I don't care. Any number will do because of the *constant multiple rule for integrals*. If we let c stand for any number (c stands for constant, a number that doesn't change for the duration of a calculation once we have chosen its value) and use $f(x)$ again to stand for any mathematical expression we care to choose, then the following calculus equation presents the constant multiple rule for integrals:

$$\int_a^b c \cdot f(x) = c \cdot \int_a^b f(x).$$

Here's an example of the constant multiple rule for integrals in action:

$$\begin{aligned} \int_4^7 6x &= 6 \cdot \int_4^7 x \\ &= 6 \cdot \frac{1}{2}(7^2 - 4^2) \\ &= 3(33) = 99. \end{aligned}$$

See that? We pulled out the 6 (but didn't throw it away, of course) and worked out the remaining integral with the power rule for integrals.

I have another surprise for you. I hope you're not tired of our friend $\int_2^6 \left(\frac{1}{2}x + 4\right)$, because we are going to turn our attention toward it one more time. If you've been thinking about what we've done, you might very well have anticipated this result and the surprise will be ruined, but that's okay. Permit me to congratulate you on catching on to where we're going before we quite got there.

We originally evaluated $\int_2^6 \left(\frac{1}{2}x + 4\right)$ in Chapter 2 with a simple graph involving a vertical trapezoid. Then we computed it again in this chapter by breaking the integrand in two and working out each piece with separate graphs (another vertical trapezoid and a rectangle). As you'll recall, the point of that was to demonstrate the addition property of integrals. Now, however, I'm going to put the addition property together with the multiplication property, and we're going to find the value of the integral in a slick calculation involving the power rule for integrals. The point, I hope you see, is that we can get away from the graphical constructions, just pick up a pencil, and scribble out a result. Here we go:

$$
\begin{aligned}
\int_2^6 \left(\frac{1}{2}x + 4\right) &= \int_2^6 \frac{1}{2}x + \int_2^6 4 && \text{(break the integral in two)} \\
&= \frac{1}{2}\cdot\int_2^6 x + 4\cdot\int_2^6 1 && \text{(pull out the factors)} \\
&= \frac{1}{2}\cdot\frac{1}{2}(6^2 - 2^2) + 4(6-2) && \text{(apply the power rule)} \\
&= \frac{1}{4}(32) + 4(4) = 24.
\end{aligned}
$$

The result is no surprise, although we're glad it came out to be 24 again. A different result would have been an embarrassment after all this work. However, perhaps you were startled just a bit when I factored out the 4 from the second integrand and left a 1 behind. I needed to do that to conform to our power rule for integrals, which covers only integrands of the form $1, x, x^2$, and so on. Once that was done, however, the power rule took care of everything.

I think we've done almost enough work for one chapter. There is, however, one more important point to make—one that wraps up all of our work into one neat package. The rules we have just deduced for integrals mean that we can now easily integrate any polynomial. Recall that polynomials are expressions involving sums, differences, and multiples of powers of x. Some examples are $x^2 + 5x + 6$, $5x^8 - 6x^2$, and $x^3 - 4x + 4$. I've graphed the last example, $x^3 - 4x + 4$, in Figure 7-6 for

Figure 7-6.

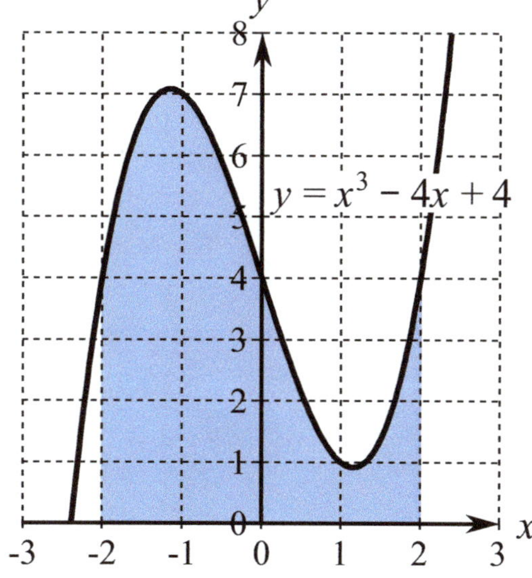

you. We can easily integrate this polynomial for $-2 \leq x \leq 2$, as follows:

$$\begin{aligned}\int_{-2}^{2}(x^3 + 4x + 4) &= \int_{-2}^{2} x^3 - 4 \cdot \int_{-2}^{2} x + 4 \cdot \int_{-2}^{2} 1 \\ &= \frac{1}{4}(2^4 - (-2)^4) - 4 \cdot \frac{1}{2}(2^2 - (-2)^2) + 4(2 - (-2)) \\ &= \frac{1}{4}(0) - 2(0) + 4(4) = 16.\end{aligned}$$

If you like, you're welcome to go to Figure 7-6 and start trying to count the squares. Does it look like 16 is a reasonable answer? (I assure you that it is.)

We have reached a major milestone. As it turns out, polynomials are extremely important in mathematics because they (a) provide good models for many problems in science and economics (for example) and (b) are extremely easy to use in calculus computations. You will see some examples of this in the next chapter.

Postscript

Okay, did you catch me? I may have slipped one by you. Although I presented you with reasons to accept the addition rule and constant multiple rules for integrals, I never said anything about a *subtraction* rule. Then, without any apology or warning, I worked out a final example that had a subtraction in it. Yes, the subtraction rule works just fine, but the pictures aren't as pretty or easy to draw. Try your hand at sketching some figures for simple examples and see if you can explain the subtraction rule to your own satisfaction. I've provided one example in Figure 7-7 as a hint. Good luck!

Figure 7-7.

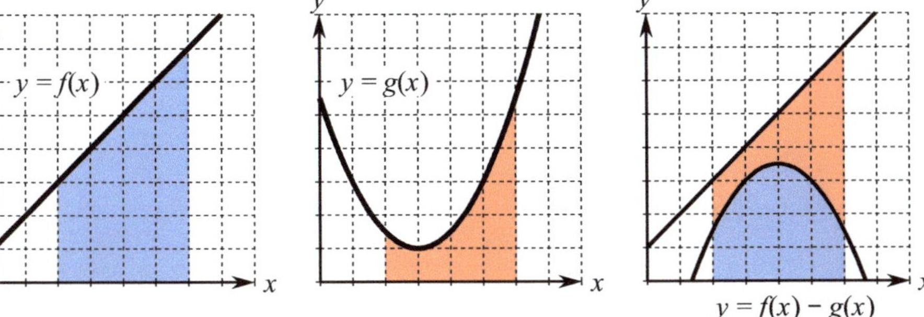

8 It Slices, Dices, and Spins

A few turns on Cavalieri's lathe

Today we are going to slice up a cone. The goal is to find its volume. As usual, it's a good idea to begin with a specific example. How about a cone whose height is 2 ft and whose circular base has a radius of 1 ft? Our example, shown in Figure 8-1, is a right circular cone. The "right" means that the height of the cone is perpendicular (at right angles) to the base. The "circular," of course, just means that the cone has circular cross-sections if we slice through it horizontally, as we are about to do.

Figure 8-1.

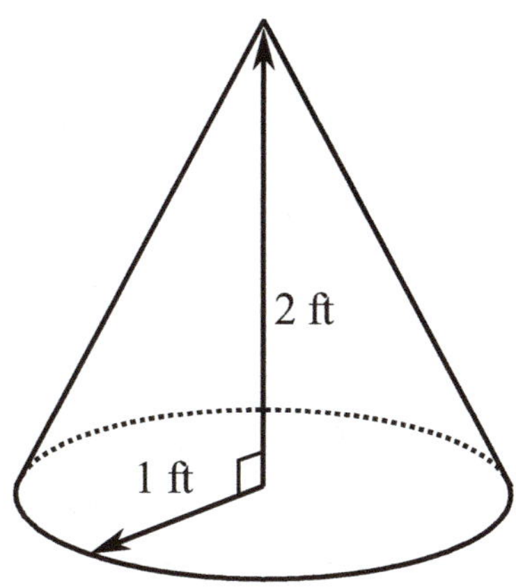

Actually, I take that back. We are going to look at circular cross-sections of the cone, but we're going to do it by slicing vertically. That's because I'm going to place the cone on its side first, as opposed to leaving it standing upright. It's a matter of convenience. In this example, I'm going to show you how to embed the cone in a coordinate system so that we can get a nice mathematical description of its cross-sections. That's going to be easier if each cross-section corresponds to a unique point on the x axis of our coordinate system. Each cross-section, as just noted, will be a circle. We know how to find the area of circles. Also recall that we can find volume by integrating area, as we did in our earliest examples in Chapter 1. That's our plan for finding the volume of our right circular cone.

Before we get too far into it, let me remind you about something very fundamental concerning the Cartesian coordinate system in which we are going to embed the cone. If you are at the point with coordinates (x, y), then you automatically know two very important distances. As shown in Figure 8-2, x is the horizontal distance of the point (x, y) from the origin (the place where the x and y axes intersect) and y is the

Figure 8-2.

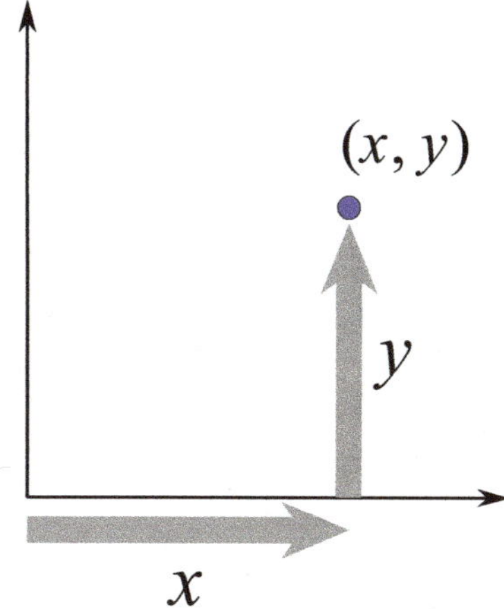

vertical distance. With that simple fact in mind, let's see how we can use those distance measures to work up a nice description of our cone.

Please examine Figure 8-3, where I have drawn the cone on top of an *xy* coordinate system. I have placed the cone so that it's lying on its side. Its vertex (the sharp point) is at the origin—the point (0, 0)—and the axis of the cone coincides with the *x* axis. I've also drawn in the line whose equation is $y = \frac{1}{2}x$. As you can see, this line passes right along the edge of the cone. We can use this fact to work out a formula that gives us the area of every circular cross-section of the cone.

Figure 8-3.

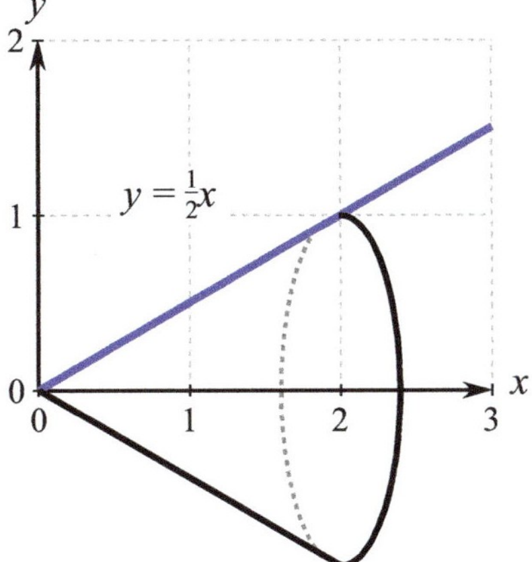

To see this, pick any value of *x* you like between 0 and 2. In Figure 8-4, I've marked the value $x = 1.25$ with a dot on the *x* axis. What is the height of the point on the line $y = \frac{1}{2}x$ that corresponds to $x = 1.25$? It's just $y = \frac{1}{2}(1.25) = 0.625$, right? As you can see, I've also marked that point on the line; it's the dot with coordinates (1.25, 0.625).

Figure 8-4.

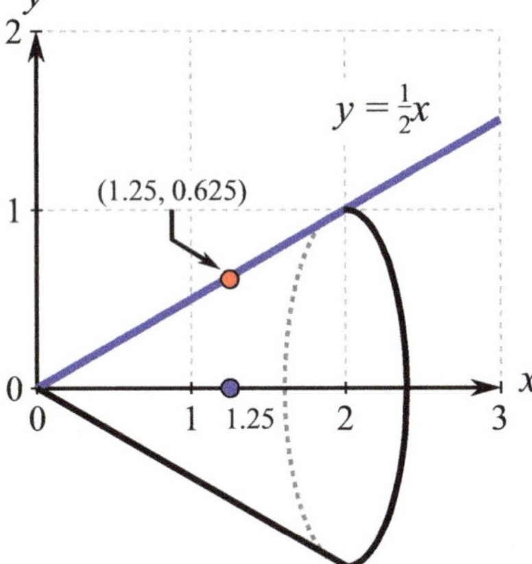

What would happen if we made a nice, neat vertical cut through the cone so that we pass through the point $x = 1.25$ on the *x* axis? If the cut is made perpendicular to the *x* axis, we would get a circular cross-section like the one shown in Figure 8-5. I've marked the radius of that cross-section with an arrow labeled *r*. What might the value of *r* be? That's no mystery: It's the value of the *y* coordinate of the point at the tip of the arrow, which we already know is 0.625. In fact, any time we choose an *x* value at which to place a cross-section, the circular cross-section is going to have a radius *r* that equals *y*, where $y = \frac{1}{2}x$.

Figure 8-5.

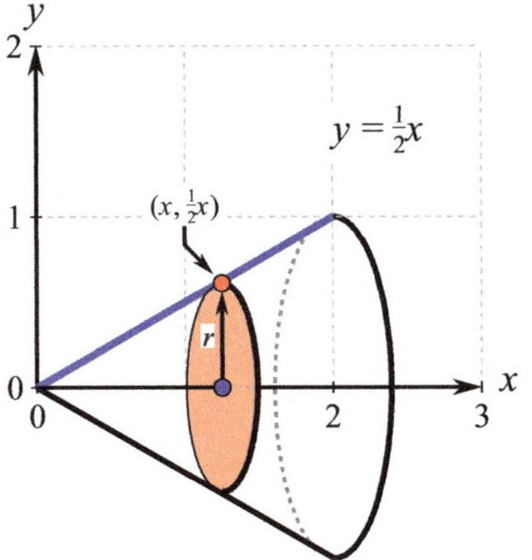

Since we know the radius of the cross-section at $x = 1.25$, we also know its area. That's because the area of any circle is just πr^2. We therefore have an area of $\pi(0.625)^2 = 0.390625\pi \approx 1.22718$ ft^2. Okay? That was all for the case where $x = 1.25$. We have used a very specific example to set up the approach we need to construct a more general result, which is where we turn our attention next.

What if we don't say what x is? What if we compute the result generically, without specifying the x value? No problem. Retracing our steps, using the generic symbol x wherever we previously used 1.25, we are still going to end up with a circular cross-section whose radius is just y again. (See Figure 8-6.) This time we don't have a specific numerical value for y, but that's okay. We know that $y = \frac{1}{2}x$ in this example, so we can compute the area very easily:

$$\pi r^2 = \pi y^2 = \pi\left(\frac{1}{2}x\right)^2 = \pi\left(\frac{1}{4}x^2\right) = \frac{\pi}{4}x^2.$$

(Use your calculator to plug the specific value $x = 1.25$ into this generic result. You should, of course, get 1.22718 again, just as we did before. Did it work?)

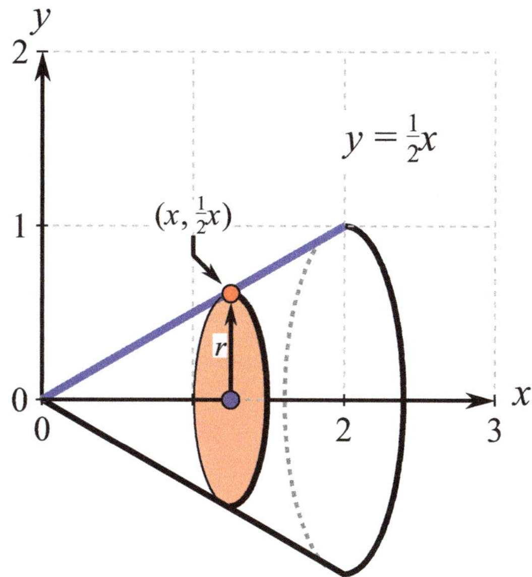

Figure 8-6.

We have a brand-new expression now. This new expression provides a formula for the area of each circular cross-section of our cone. Using A for area, I'll write down our new expression as

$$A = \frac{\pi}{4}x^2.$$

This expression is not too difficult to graph, as you can see from Figure 8-7. I shaded the area under the curve for x values between 0 and 2 because that area corresponds to the range of values for the circular cross-sections of our cone. (Remember the cone we started with?) If we integrate an area expression, we get a volume. So now we want to know if we can find the integral of this area expression and thereby find the volume of the cone.

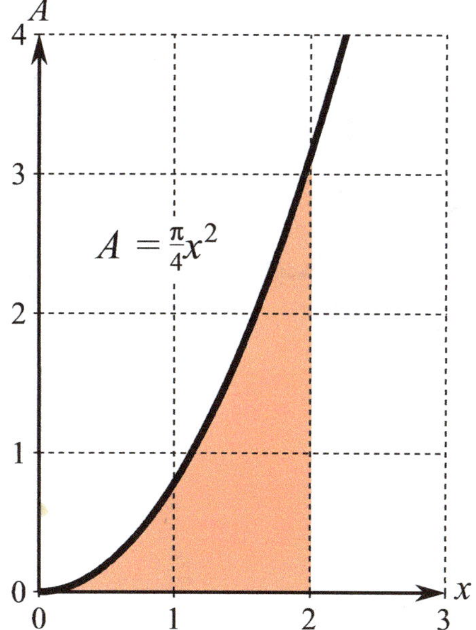

Figure 8-7.

With the assistance of our integration rules, it's not difficult at all:

$$\text{Cone volume} = \int_0^2 \frac{\pi}{4}x^2 = \frac{\pi}{4}\int_0^2 x^2 = \frac{\pi}{4}\cdot\frac{1}{3}(2^3 - 0^3) = \frac{\pi}{4}\cdot\frac{1}{3}\cdot 8 = \frac{2\pi}{3}.$$

Since our original units of measurement were feet, this result should be expressed in cubic feet: $\frac{2\pi}{3}$ ft^3. We have found the volume of our cone, a right circular cone whose height was 2 ft and whose circular base had a radius of 1 ft.

How about finding the volume of a more general cone, say, one with height h and radius a? If we embed the cone in a coordinate system, as we did before, with the vertex at the origin and the cone's axis along the x axis (see Figure 8-8), we can easily see a formula for the area of the circular cross-sections. The line $y = \frac{a}{h}x$ coincides with the upper edge of the cone, so the cross-sectional area corresponding to any particular choice of x will be

$$A = \pi y^2 = \pi\left(\frac{a}{h}x\right)^2 = \frac{\pi a^2}{h^2}x^2.$$

If we integrate this area expression to obtain the volume of our new cone, we have

$$\text{Volume of cone} = \int_0^h \frac{\pi a^2}{h^2}x^2 = \frac{\pi a^2}{h^2}\cdot\frac{1}{3}(h^3 - 0^3) = \frac{\pi a^2 h}{3}.$$

If you check any algebra or geometry textbook, this is the formula you'll find for the volume of a right circular cone. We worked it out ourselves by integrating the cone's cross-sectional area.

Figure 8-8.

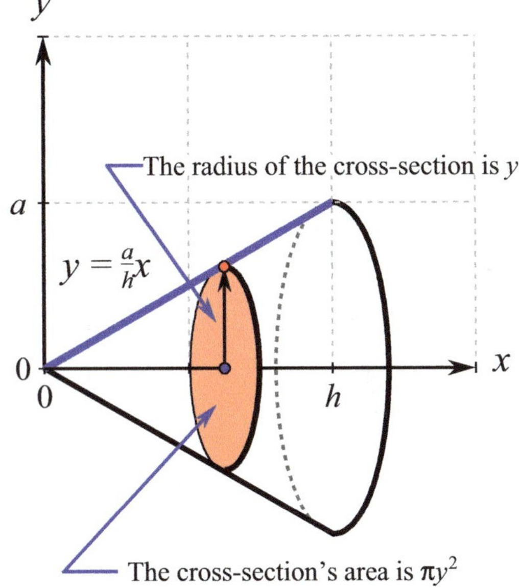

Let us now apply what we learned in computing the volumes of the cones to the calculation of the volume of a different solid. Consider a sphere of radius a. You may already know that the volume of a sphere of radius a is $\frac{4}{3}\pi a^3$, but we're going to work it out for ourselves. The procedure will be the same as for the cones: embed the sphere in a coordinate system, find an expression for the sphere's cross-sectional areas, and integrate that area expression.

Figure 8-9.

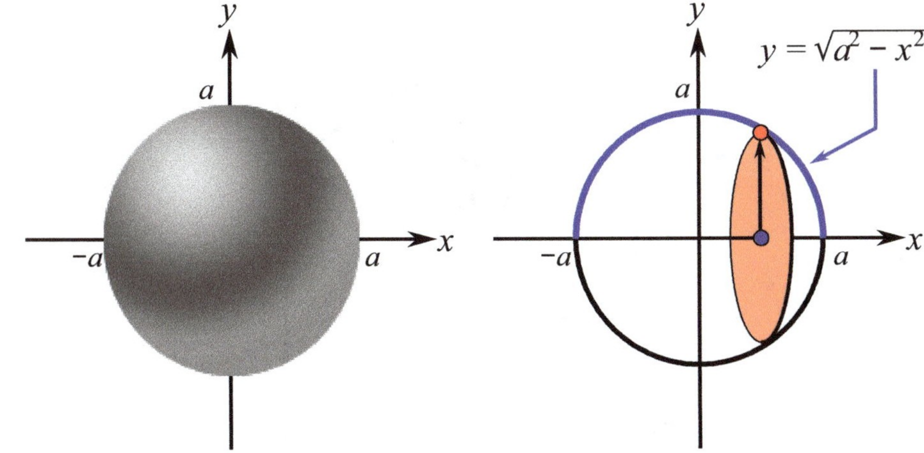

The left side of Figure 8-9 shows the sphere positioned with its center at the center of a rectangular coordinate system. The right side of Figure 8-9 shows the semicircular curve in the coordinate grid that coincides with the top of the sphere. It has a formula you should recall:

$y = \sqrt{a^2 - x^2}$. (Do you remember learning in algebra that $x^2 + y^2 = a^2$ is the equation of a circle of radius a? When you solve this equation for y, you get the square root formula I just gave.)

Just as with our calculation for the cone, we compute the cross-sectional area for the sphere by working out πy^2:

$$A = \pi y^2 = \pi(\sqrt{a^2 - x^2})^2 = \pi(a^2 - x^2).$$

As we see from Figure 8-8, we need to integrate this area expression for x values between $-a$ and a. The result is

$$\text{Volume of sphere} = \int_{-a}^{a} \pi(a^2 - x^2) = \int_{-a}^{a} \pi a^2 - \int_{-a}^{a} \pi x^2$$
$$= \pi a^2 \int_{-a}^{a} 1 - \pi \int_{-a}^{a} x^2 = \pi a^2[a - (-a)] - \pi \cdot \frac{1}{3}[a^3 - (-a)^3]$$
$$= \pi a^2[2a] - \pi \cdot \frac{1}{3}[2a^3] = 2\pi a^3 - \frac{2}{3}\pi a^3 = \frac{4}{3}\pi a^3.$$

See? We got the expected result.

The technique I've outlined in this chapter can be applied to find the volume of any suitable object. What makes an object suitable for this sort of treatment? It needs to be what we call a *solid of revolution*; that is, any object that could have been produced by turning on a lathe (see Figure 8-10). No matter how you spin it, it looks the same. This is certainly true of the cone and the sphere.

A key step in computing the volumes of our examples involved finding a suitable mathematical expression to describe the region's boundary after we embedded it in a coordinate system. For our first cone, the boundary expression was $y = \frac{1}{2}x$. For our second, it was $y = \frac{a}{h}x$. For the sphere, it was $y = \sqrt{a^2 - x^2}$. Suppose in some general case it was given by $y = f(x)$, where $f(x)$ is some generic mathematical function to be specified later. Then our area expression will once again be $A = \pi y^2 = \pi[f(x)]^2$.

Figure 8-10.

Courtesy of Missouri School for the Deaf, www.msd.k12.mo.us

(See Figure 8-11.) If the math function $f(x)$ is defined over the range of x values from a to b, then the volume of the corresponding solid of revolution would be given by

$$\text{Volume} = \int_a^b \pi[f(x)]^2.$$

This, then, is the general formula that mathematicians and scientists use to compute the volume of a solid of revolution. Now we can do it, too, provided that we can evaluate the integral.

Figure 8-11.

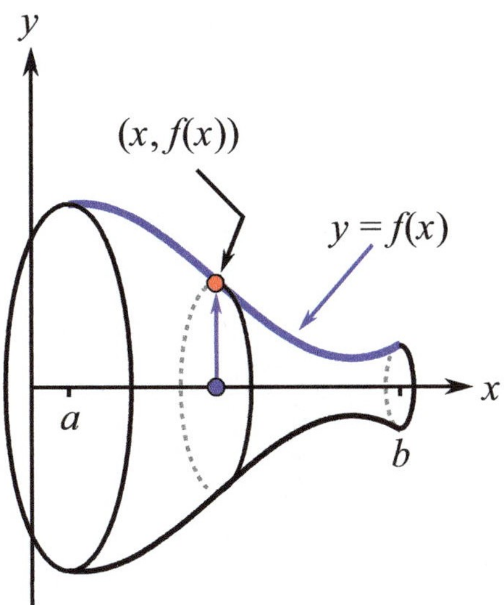

9 In the General Area

Moving the goal posts

There has been a general method to my madness throughout our leisurely stroll down the calculus path. I like to take us through stages, beginning with a very specific example, then redoing the example in a different way (to reinforce what we just did), and finally trying to present a more general example or technique that we will be able to use later. You will recognize this as a typical teaching method, moving from the specific to the general. Mathematicians are particularly fond of general formulations, results that solve as broad a range of problems as possible, leaving out only the special details that anyone can plug in later.

Here's a simple reminder of what I mean: If I were to tell you that an object moving at a velocity of 4 miles per hour will travel $y = 4t$ miles in t hours, you would quickly see that this formula is true. It's a great formula—provided we're talking about objects that travel 4 mi/hr. But you already know a much better formula: An object moving at a velocity of r for a period of time t will travel $d = rt$ units of distance. It's our old friend "Distance equals Rate times Time," expressed in perfect generality. I didn't even identify any of the units of measurement. As long as we make consistent choices, we could be talking about miles per hour and hours, meters per second and seconds, or furlongs per fortnight and fortnights (a particular favorite of undergraduate physics majors with too much time on their hands). Now that's what I mean by a general formula.

In this chapter, I'm going to show you how to generalize integrals a little bit. In most of our examples to date, we've used limits of integration like 0 and 1, 2 and 6, or even a and b. In each case, though, it's been understood that the limits of integration are constants, not

variables. (Even in the case of a and b, the letters are just placeholders for constants to be specified later.) There was, however, an exception. I presented this example back when we were first talking about representing distance traveled as the area under a graph. Here it is again, recycled from Chapter 3:

An object traveling with constant acceleration has the velocity $y = at$, where a is a constant and t stands for time. The distance traveled during constant acceleration during the time interval from 0 to t will be the area bounded above by the line $y =$ at, as shown in Figure 9-1. We find this distance by computing the integral of $y = at$:

$$\begin{aligned}\int_0^t at &= a \cdot \int_0^t t \\ &= a \cdot \frac{1}{2}(t^2 - 0^2) \\ &= \frac{1}{2}at^2.\end{aligned}$$

Once again we have the famous $d = \frac{1}{2}at^2$ formula for distance traveled at constant acceleration, a result known to Galileo. We did it by allowing t, a variable, to be the upper limit of integration.

Figure 9-1.

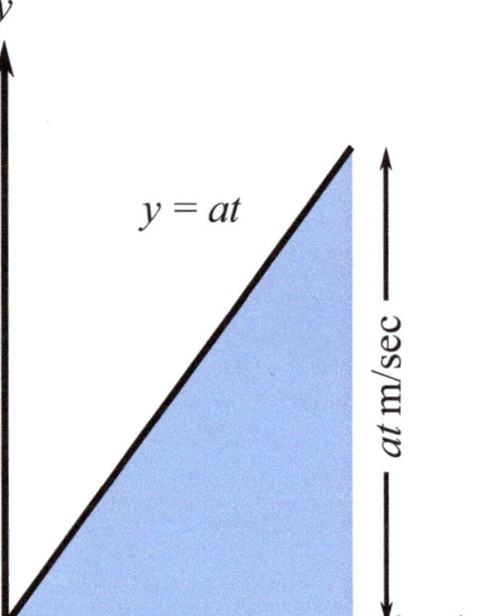

When we originally computed this result in Chapter 3, we used the area of a triangle instead of the power rule for integrals (which we didn't know yet), but I didn't recycle this example just to show you another way to work the calculation. My reason for bringing this example back is to show you that we can compute integrals whose limits of integration are *variables*. Ordinarily, when we compute an integral, the result is a number, like 8, or $\frac{2\pi}{3}$, or something like that. Now we're going to talk about integrals with variable limits and variable results, because that will set the stage for the next big step in our calculus journey. We won't be able to tackle the famous Fundamental Theorem of Calculus (still a few chapters away) until we generalize the integral in this fashion. Since we have a head start, it won't take too long to complete this leg of our trip.

We're going to use the notation $A(x)$ to represent the value of an integral whose upper limit of integration is the variable x. The letter A was chosen to remind us of "area," although we know that the result of an integral is not limited to that interpretation. The notation $A(x)$ is, of course, in the form of function notation, which we've already used on a few occasions when we needed $f(x)$ and $g(x)$ to stand in for unspecified mathematical expressions. Let's work out an actual example by computing $A(x)$ in the following instance:

$$A(x) = \int_0^x x^2 = \frac{1}{3}(x^3 - 0^3) = \frac{1}{3}x^3.$$

In this example, $A(x)$ is a handy formula for the area under the parabola $y = x^2$ for the region from 0 to x, as shown in Figure 9-2.

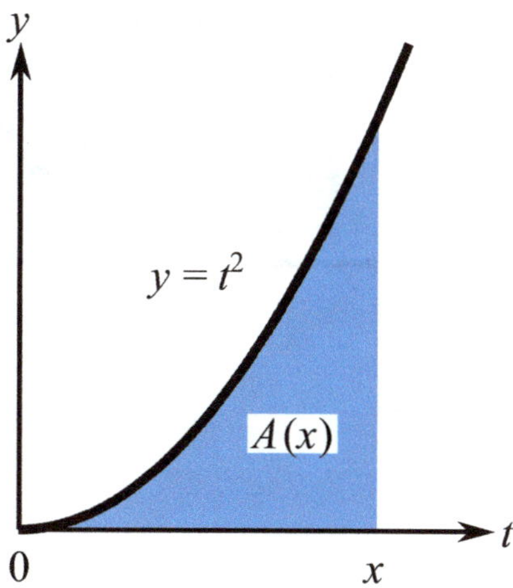

Figure 9-2.

You may have already noticed the rather ambiguous role of x in the integral we just computed. It is both in the integrand and in the upper limit of integration. I should mention that this is considered bad form by most mathematicians. There is an extremely simple way to avoid this minor problem. Consider this computation:

$$A(x) = \int_0^x t^2 = \frac{1}{3}(x^3 - 0^3) = \frac{1}{3}x^3.$$

I wrote the integrand in terms of t instead of x. The result, however, depends on the limits of integration—the symbols that are plugged in for the final steps—and not on the variable used to represent the integrand. Since the variable used to represent the integrand is completely immaterial—it's not going to be in the final answer anyway—we call it a "dummy variable." My favorite choice of dummy variable is t, which I'll use whenever we want to restrict the use of x to the limits of integration. Let's practice a bit more with this notation, applying it to some rather familiar examples:

$$\int_0^x t = \frac{1}{2}(x^2 - 0^2) = \frac{1}{2}x^2.$$

Although it's more general now, this integral is an old acquaintance. As you can see from Figure 9-3, we can easily verify the result because it's the area of a triangular region. Even simpler is the region in Figure 9-4, which corresponds to the integral

$$\int_0^x 1 = x - 0 = x.$$

Whether we write these integrands as $y = 1$, $y = t$, and $y = t^2$, or $y = 1$, $y = x$, and $y = x^2$, as we did before, their integrals are easy to compute with the formulas we've developed so far. Nor does it matter whether the limits of integration are constants like 0 and 1 (or implied constants like a and b) or include a variable (as with 0 and x)—the power rule still shows us what to do.

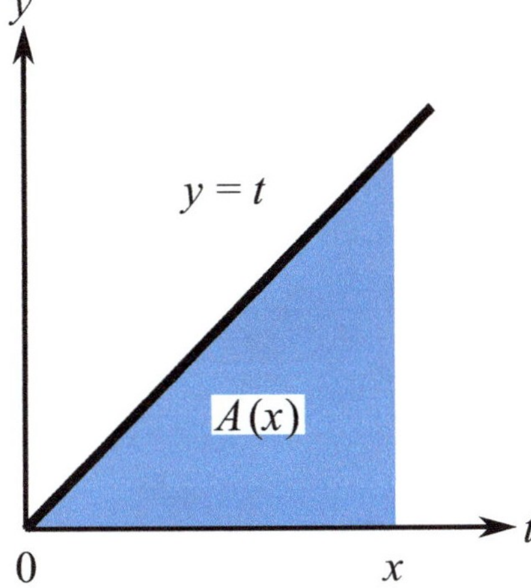

Figure 9-3.

I want to introduce one more variation on the theme on turning integrals into expressions with variables in them. The lower limit of integration doesn't have to be 0. Consider Figure 9-5. Here's an example that defines $A(x)$ for a region that ranges from 1 to x instead:

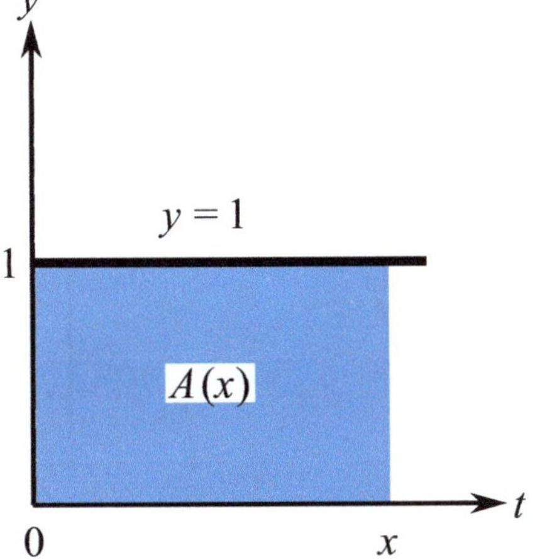

Figure 9-4.

$$\begin{aligned} A(x) &= \int_1^x (4t - t^2) \\ &= 4 \cdot \frac{1}{2}(x^2 - 1^2) - \frac{1}{3}(x^3 - 1^3) \\ &= 2(x^2 - 1) - \frac{1}{3}(x^3 - 1) \\ &= 2x^2 - 2 - \frac{1}{3}x^3 + \frac{1}{3} = 2x^2 - \frac{1}{3}x^3 - \frac{5}{3}. \end{aligned}$$

We may now choose any value of x between 1 and 4, plug it into our result for $A(x)$, and quickly obtain the measure of the corresponding area underneath the parabola. The formula for $A(x)$ provides a more general result than most of our previous computations.

We have now turned integrals into functions that depend on a variable. I mentioned earlier that it would be useful to do this because it creates more general results. There's another reason, too. Calculus includes some powerful tools to analyze the behavior of functions. I am about to introduce you to a few of those tools in the next chapter, after which we will apply them to integrals. That, in turn, will take us to the crucial theorem that lies at the heart of calculus.

Figure 9-5.

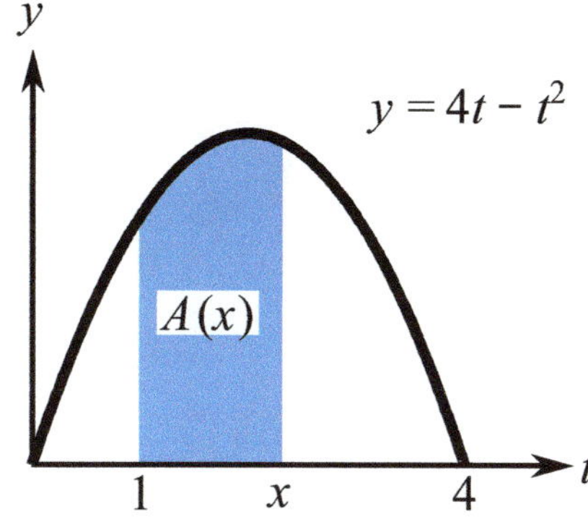

10 Hitting the Slopes

A steep inclination

We really are going off in a new direction in this chapter. Our attention has been fixed on just half of the calculus story. I am going to introduce you to the other half now.

Since calculus was developed by people who were interested in solving real problems in science—particularly physics—it should be no surprise that many of our examples come directly from scientific applications. We have worked out several problems in which computing the area under a velocity curve gave us a distance measure. What if someone asked us to solve the opposite problem? That is, what if you already knew the distance traveled during each interval of time and wanted to know the velocity? While it obviously should be related to what we did before, it is certainly a different kind of computation.

It often helps, when you're looking at a new kind of problem, to consider the simplest possible case. We began in Chapter 1 by looking at $d = rt$ (Distance equals Rate times Time) as represented by the area of a rectangle. Let's take a moment to consider what we would do if we were given the distance traveled and the time interval over which the travel occurred. How could you then find the rate (the velocity)? It's pretty easy, right?

For a specific example (you can see I'm dredging up the one we started with), suppose that an object traveled 12 feet in 4 seconds. We can plug into our formula for distance traveled and solve for the remaining quantity:

$$
\begin{aligned}
d &= rt \\
12\text{ ft} &= r \cdot 4\text{ sec} \quad &&\text{(substitute what we were given)} \\
\frac{12\text{ ft}}{4\text{ sec}} &= r &&\text{(divide both sides by 4 sec)} \\
3\frac{\text{ft}}{\text{sec}} &= r. &&\text{(simplify)}
\end{aligned}
$$

Now we know that an object traveling at a constant velocity must go 3 ft/sec in order to cover a distance of 12 feet in 4 seconds. Big surprise, right? Don't worry. It's about to get a little more interesting and challenging.

Let me draw you a nice graph of the journey that corresponds to our introductory example. The object is moving along at a rate of $r = 3$ ft/sec. Click a stopwatch to track its motion with respect to time. At the initial click, when $t = 0$ seconds, the object hasn't had a chance to go anywhere, so the initial distance traveled is $d = 0$ feet. After $t = 1$ second of time, however, the object has moved 3 feet (3 feet per second, right?). Then after $t = 2$ seconds, it's at $d = 6$ feet; $t = 3$ seconds, $d = 9$ feet; and $t = 4$ seconds, $d = 12$ feet. Let's record this information in a distance versus time graph. The horizontal axis will be t for time and the vertical axis d for distance. The result is Figure 10-1.

Figure 10-1.

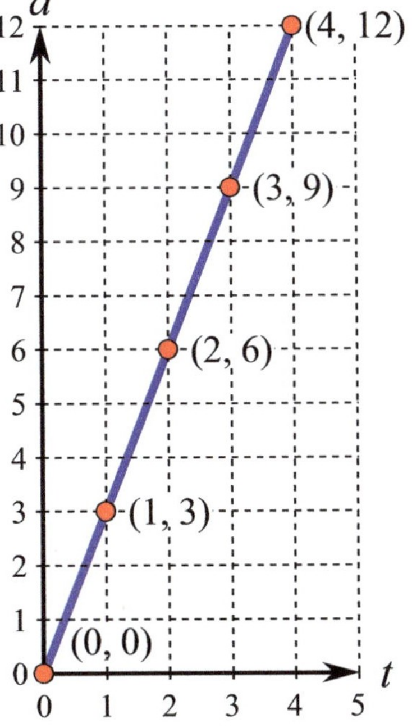

I have a question for you. You can see where time is recorded in the graph (it's the horizontal axis). You can see where distance is recorded in the graph (it's the vertical axis). *Where is the rate?* Can you find anything that looks like 3 feet per second? This is the key question. Think about it for a minute before you read on. I'm serious. Think about it for a while and look at the graph. Try to find 3 in there. (And the 3's on the axes and the 3's in (1, 3) and (3, 9) don't count!)

If reading this book has stirred your recollection of algebra, you probably know where the 3 is. It's recorded in the *steepness* of the graph. If the object were traveling faster, then the distance would grow more rapidly. Since distance is recorded as the vertical coordinate in our graph, that means the line would have to rise up more rapidly. It would become steeper. Conversely, a slower object would have a graph containing a line that would rise more slowly; it wouldn't be as steep.

As it turns out, the word "steep" has not caught on in mathematical circles. We prefer the word "slope." For the graph in Figure 10-1, the slope is 3 feet per second. You find it by computing a ratio: the quotient of the vertical change in the graph and the horizontal change in the graph. Since the vertical change was 12 feet and the horizontal change was 4 seconds, the slope is

$$\frac{12\text{ ft}}{4\text{ sec}} = 3\text{ ft/sec},$$

a quantity we had already computed earlier, but now we know where it fits into the graph.

It turns out that 3 ft/sec is *everywhere* in the graph of the line. When I used 12 ft and 4 sec, I was using the total distance and the total time. That was not necessary. I could have used any corresponding intervals of distance and time. For example, look at the points (2, 6) and (3, 9) and remember what they represent. How far did the object travel during the time interval from $t = 2$ sec to $t = 3$ sec? It was the difference between 6 ft and 9 ft:

$$9\text{ ft} - 6\text{ ft} = 3\text{ ft}.$$

How long did that time interval last? It lasted the difference between 2 sec and 3 sec:

$$3\text{ sec} - 2\text{ sec} = 1\text{ sec}.$$

Then what is the ratio of change in distance and change in time? It's

$$\frac{3\text{ ft}}{1\text{ sec}} = 3\text{ ft/sec},$$

which is just what we got before.

The usual name for the vertical change in a graph is *rise*, and the usual name for the horizontal change is *run*. That's why in algebra we say that the slope equals "rise over run." In the example we just discussed, the line in Figure 10-1 could have been expressed as $d = 3t$, where you see that the slope turns out to be the coefficient (that is, the number in front) of t. This happens every time we have a *linear equation*. That's an equation whose graph is a straight line, as in our example.

A quick review of lines

We have already been using linear equations in our examples. For instance, we've seen $y = x$, $y = 2t$, and $y = \frac{1}{2}x + 4$, among others. A linear equation can be written in the form $y = mx + b$, where m is the slope (as we just discussed) and b is the y intercept, the place where the line crosses the y axis. (Too bad that we're accustomed to using b for other things. Our most popular letters get to stand for many different things in math, and you have to keep in mind the context if you're going to know which interpretation is the correct one.)

If we look at $y = \frac{1}{2}x + 4$ again, we can see that its slope should be $\frac{1}{2}$ and its y intercept should be 4. Looking at its graph in Figure 10-2, we can verify both of these facts. That is, the line rises 1 unit for every 2 units it runs horizontally. Also, the line intersects the y axis at 4.

Figure 10-2.

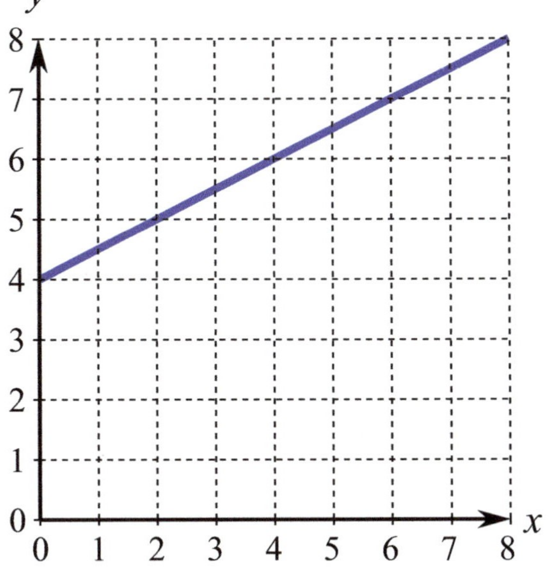

Instead of puzzling out the slope by counting squares on the graph, we can also work it out by calculation. It is common in math to use Δ, the upper-case Greek letter delta (which corresponds to the Roman letter D), to represent change or "difference." Thus Δy means "change in y" and Δx means "change in x." (Don't confuse these with products: Δy *does not* mean Δ times y.) Thus Δy and Δx stand for rise and run, respectively. If we have a pair of points lying on a particular line, we can compute Δy and Δx without difficulty; they're just differences. Suppose our two points are (x_1, y_1) and (x_2, y_2). (The subscripts merely distinguish point 1 from point 2; unlike superscripts, which indicate powers, subscripts are simply labels.) Then we can write that $\Delta y = y_2 - y_1$ and $\Delta x = x_2 - x_1$. The

formula for slope becomes a familiar old friend from algebra:

$$\text{Slope} = \frac{\text{rise}}{\text{run}} = \frac{\Delta y}{\Delta x} = \frac{y_2 - y_1}{x_2 - x_1}.$$

Let's try our formula out on the line $y = \frac{1}{2}x + 4$, whose graph is shown in Figure 10-3. We need to pick two points that lie right on the line. Do you agree that (2, 5) and (6, 7) fit the bill? (There are plenty of other choices, but we need only two.) Let's label our points so that $(x_1, y_1) = (2, 5)$ and $(x_2, y_2) = (6, 7)$. Then the rise will be $y_2 - y_1 = 7 - 5 = 2$ and the run will be $x_2 - x_1 = 6 - 2 = 4$. If we plug these values into the slope formula, we get

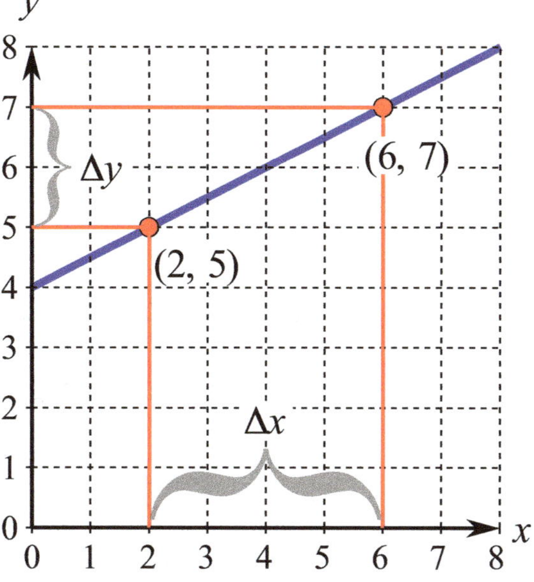

Figure 10-3.

$$\begin{aligned}\text{Slope} &= \frac{\text{rise}}{\text{run}} \\ &= \frac{\Delta y}{\Delta x} = \frac{y_2 - y_1}{x_2 - x_1} \\ &= \frac{7-5}{6-2} = \frac{2}{4} = \frac{1}{2}.\end{aligned}$$

As we expected, the slope turned out to be $\frac{1}{2}$.

Raising the bar

We just looked at a problem that was the opposite of what we had been doing before. We discovered that it led us directly into a review of some algebra facts about lines and their slopes. I suppose we could be forgiven for thinking that we made short work of a potentially difficult problem, but we need to remind ourselves that we focused on a very simple case.

We were looking for a rate of change (velocity) and found out we could get what we wanted by computing the ratio known as slope. Unfortunately, we know how to compute slopes only for linear functions! The moment our situation gets more complicated, we won't be able to do that anymore. Let's take the problem up just one notch. Consider the parabola $y = x^2$. (Recall that this is what we call a *degree*

2 or *quadratic* function, as opposed to a *degree 1* or *linear* function.) We see in Figure 10-4 that the graph of $y = x^2$ does not have a slope. Or maybe it's better to say that it has too many! Try computing slope using points on the curve. I've marked (1, 1), (2, 4), and (3, 9) for you. If you check (and I think you should), you'll discover that the first two points give you a slope of 3 and the last two points give instead a slope of 5. If you pick other pairs of points—there are lots to choose from!—you'll get many other slope values. The parabola's steepness is always changing from place to place, whereas a line has the same slope for its entire length. The rate of change problem just became much more complicated.

The slope (whatever that means) of $y = x^2$ is the first order of business in the next chapter.

Figure 10-4.

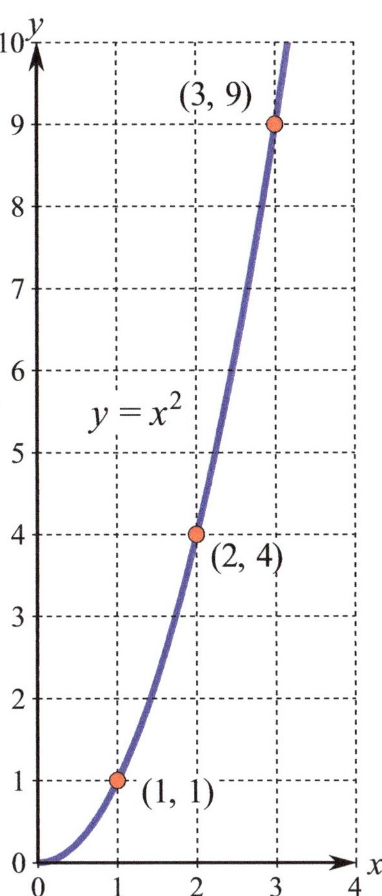

11 Off on a Tangent

Take it to the limit one more time

Can we use "slope" in a way that makes sense when we're talking about the parabola $y = x^2$? It seems as if we should. After all, slope is a measure of change, and no one can deny that the curve of the parabola changes even more dramatically than the straight-line graph of a linear function. Of course, the big advantage of the straight line is that it changes at a constant rate: one number suffices to describe its rate of change.

Obviously this implies that "slope" for a parabolic curve must be different in different places. One number can't do the trick. Just to narrow things down for starters, let's talk about what we might mean by the slope of the parabola $y = x^2$ at the point (1, 1). We normally need *two* points to compute a slope. What if I pick (2, 4) for my second point? Then we would have a slope equal to

$$m = \frac{\text{rise}}{\text{run}} = \frac{y_2 - y_1}{x_2 - x_1} = \frac{4 - 1}{2 - 1} = \frac{3}{1} = 3.$$

This computation has the advantage of including the point (1, 1), the point we're interested in, but is (2, 4) really a good choice for the second point? In Figure 11-1, you can see where I drew in a line connecting (1, 1) and (2, 4). That line is certainly close to the parabolic arc, but it's clearly not the same. Let's try to improve things. What if I chose to use (1.5, 2.25) as my second point? Wouldn't that be better? Now we get

$$m = \frac{\text{rise}}{\text{run}} = \frac{y_2 - y_1}{x_2 - x_1} = \frac{2.25 - 1}{1.5 - 1} = \frac{1.25}{0.5} = 2.5.$$

This calculation gave us a smaller result. Is that what's going to happen every time we pick a point closer to (1, 1)? Let's try (1.1, 1.21):

Figure 11-1.

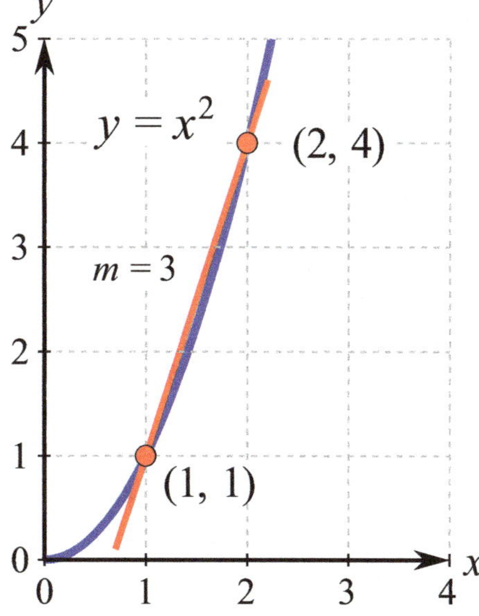

$$m = \frac{\text{rise}}{\text{run}} = \frac{y_2 - y_1}{x_2 - x_1} = \frac{1.21 - 1}{1.1 - 1} = \frac{0.21}{0.1} = 2.1.$$

Yes, it happened again. As you can see in Figure 11-2, each time we chose a point closer to (1, 1) with which to compute a slope, the value of the slope dropped. How far can this continue? Let's use (1.01, 1.0201):

$$m = \frac{\text{rise}}{\text{run}} = \frac{y_2 - y_1}{x_2 - x_1} = \frac{1.0201 - 1}{1.01 - 1} = \frac{0.0201}{0.01} = 2.01.$$

Is the picture getting clearer? It looks like our slope value is edging ever closer to $m = 2$. Do we want to say that this is the slope at the point (1, 1)? We can't actually use (1, 1) for both the first point and the second point, because that would give us a ratio of zero over zero, which is undefined. Yet everything short of that is just an approximation to what is happening *at* (1, 1). How can we settle this?

Figure 11-2.

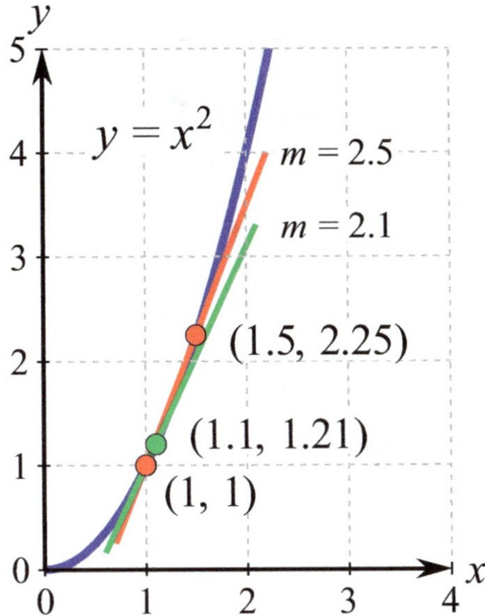

The secret is a mathematical tool called a *limit*. Mathematical limits were invented especially for calculus computations, and it's time to get acquainted with them. First, we need to generalize our set-up so that it's suitable for what I'm going to show you.

Remembering that generality is one of our most powerful techniques, let's start over with $y = x^2$ at the point (1, 1), but this time we'll pick a *generic* point with which to compute our slope. To represent our second point, I want the x coordinate to be $1 + \Delta x$. If we assume that Δx is positive, then $1 + \Delta x$ is greater than (to the right of) 1. Since our curve is the parabola $y = x^2$, the corresponding y value has to be $(1 + \Delta x)^2$. We can repeat our slope calculation, this time using the points (1, 1) and $(1 + \Delta x, (1 + \Delta x)^2)$:

$$m = \frac{y_2 - y_1}{x_2 - x_1} = \frac{(1 + \Delta x)^2 - 1}{(1 + \Delta x) - 1}.$$

This computation will not be as simple as the previous ones. We need some algebra. In the numerator, we're going to replace $(1 + \Delta x)^2$ with

$1 + 2\Delta x + \Delta x^2$, which is easily done with the FOIL technique (see the sidebar). We now have

$$\begin{aligned} m &= \frac{(1+\Delta x)^2 - 1}{(1+\Delta x) - 1} \\ &= \frac{1 + 2\Delta x + \Delta x^2 - 1}{1 + \Delta x - 1} \\ &= \frac{2\Delta x + \Delta x^2}{\Delta x}. \end{aligned}$$

Now we notice that the two terms in the numerator have a common factor of Δx, which we can factor out and reduce:

$$\begin{aligned} m &= \frac{2\Delta x + \Delta x^2}{\Delta x} \\ &= \frac{\Delta x(2 + \Delta x)}{\Delta x} \\ &= 2 + \Delta x. \end{aligned}$$

The slope between the points (1, 1) and $(1 + \Delta x, (1 + \Delta x)^2)$ is given by $m = 2 + \Delta x$. If we look back at our earlier calculations and consider the case where $\Delta x = 0.1$, that corresponds to the time we used the point (1.1, 1.21) as our second point and obtained $m = 2.1$ as our slope. This agrees with our new formula $m = 2 + \Delta x$.

Now comes the moment of truth: How small can Δx be? As long as it is not actually zero, the points (1, 1) and $(1 + \Delta x, (1 + \Delta x)^2)$ will be two distinct points and we can use them to compute a slope (which turns out to be $2 + \Delta x$). But Δx can be *as close to 0 as we want*, so the question becomes not "What is the value?" but "What is the ultimate or bounding value?" Newton actually used language like this when he referred to "ultimate ratios" while working on inventing calculus. Today we call the ultimate value a *limit*, which was not a term Newton used, and we have notation to represent our conclusion. We say that the limit of $2 + \Delta x$ as Δx *approaches* 0 is 2. (The limit *is* 2, although Δx is understood merely

FOILed again

The FOIL technique is a way to multiply a pair of two-term polynomials together:

$$(a + b)(c + d) = ac + ad + bc + bd$$

where *ac* is the product of the First terms, *ad* is the product of the Outer terms, *bc* is the product of the Inner terms, and *bd* is the product of the Last terms.

to *approach* 0, not equal it.) We record this information in the calculus equation

$$\lim_{\Delta x \to 0}(2 + \Delta x) = 2.$$

Later we'll see some limits that are significantly more difficult to evaluate, but we're fortunate that many of the limits of importance in calculus are intuitively easy. (If they hadn't been, progress in the early days of calculus would have been seriously held up while the first researchers puzzled out some form of limit theory.) Let me show you what I mean by "intuitively easy." What do you think happens to $5x$ as x approaches 2? That's right: $5x$ approaches 10. We write it this way:

$$\lim_{x \to 2} 5x = 5 \cdot 2 = 10.$$

We'll be working with a lot of limits that are "plug-in" limits. Sometimes, however, we need to do a fair amount of algebraic simplification before a limit can be reduced to a convenient "plug-in," as some further examples will show.

By the way, what does it mean that we just discovered the "limiting value" of the slope at (1, 1) is $m = 2$? One nice way to look at it is to draw a line with slope 2 right at (1, 1). As you can see in Figure 11-3, this is the line that touches the parabola at (1, 1). It doesn't touch the curve anywhere else in the vicinity. We call this the *tangent line* to the parabola at the point (1, 1). Mathematicians like to think of the tangent line at a point as the straight line that most closely approximates the curve in the vicinity of that point. You can already see from Figure 11-3 how good a match the tangent line is with the parabola—at least if you're close to (1, 1). In fact, if you zoomed in closer, you soon couldn't even tell the difference. What's more, we now have a perfectly good way to talk about the slope of the parabola:

> **The slope at any point of a curve is defined to be the slope of the tangent line at that point.**

Figure 11-3.

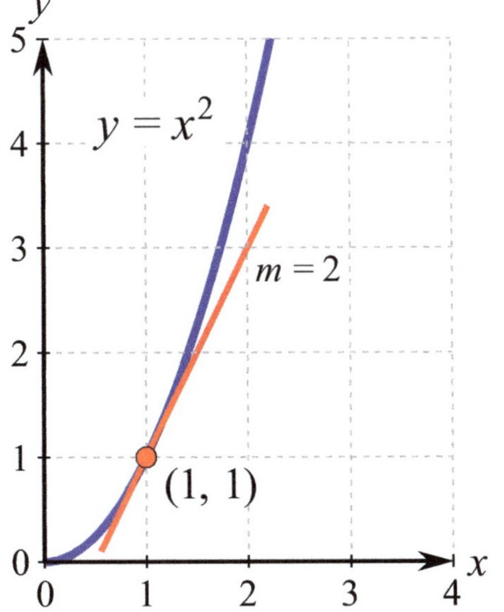

I did not specifically mention a parabola in this definition because we will apply it to any curve, not just parabolas that arise from quadratic (second-degree) polynomials.

How can we interpret this in terms of a physical model? Let's think about an object that travels a distance given by $d = t^2$ feet in t seconds. What would be its velocity at time $t = 1$ second? At $t = 1$ second, the object's position would be $d = 1^2 = 1$ foot from its starting point. That's why (1, 1) is part of the graph. If we graphed the trip according to the formula $d = t^2$, we get Figure 11-4, which is just a slightly relabeled version of Figure 11-3. I left in the tangent line at (1, 1), whose slope is $m = 2$. Actually, we can do better than that. Given the units of measurement we're using, the slope is actually 2 *feet per second*. (This makes good sense because the rise in our graph is in feet and the run is in seconds.) This isn't exactly a surprise, because we found out at the beginning of the previous chapter that slope gives you velocity whenever you have a graph that plots distance versus time. Of course, at that time we were looking at graphs that were merely straight lines. Now that we can compute slopes for more complicated curves, we can figure out velocities in those cases, too.

Figure 11-4.

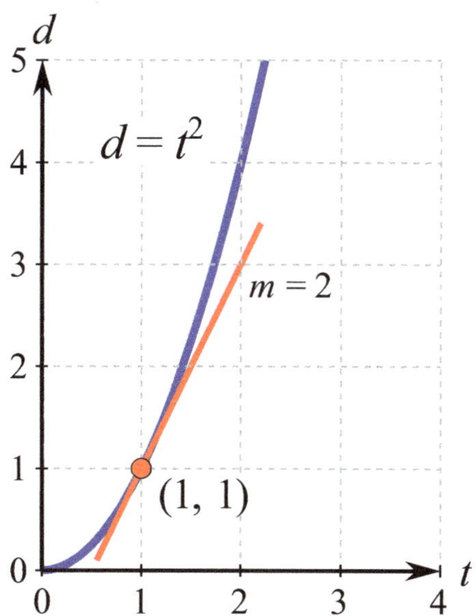

The whole parabola

To wrap up this chapter, I want to work out again our example of $y = x^2$, but this time I intend to do it generically, without picking a specific point. Instead of using (1, 1) as my starting point, I'm going to use (x, x^2). For my second point, I'm going to add Δx to x. That gives me $x + \Delta x$, whose corresponding y value has to be $(x + \Delta x)^2$, of course. You can see my set-up in Figure 11-5.

Using the points (x, x^2) and $(x + \Delta x, (x + \Delta x)^2)$, we compute the slope. After we simplify the slope expression, we find its limit as Δx goes to 0. Check the steps carefully as you read the calculation on the next page. Quite a bit of algebra is involved.

Figure 11-5.

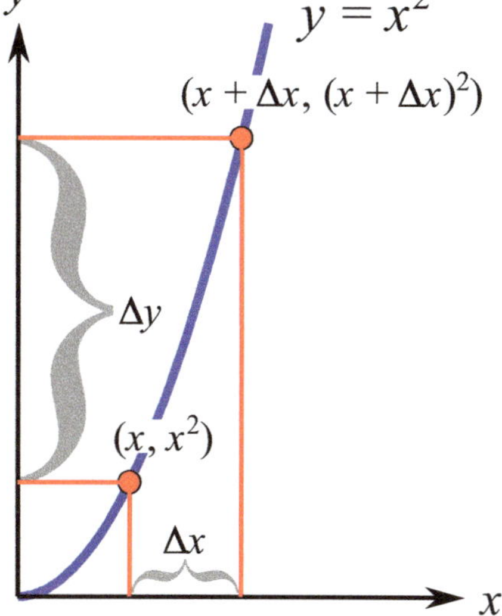

$$
\begin{aligned}
\frac{\Delta y}{\Delta x} &= \frac{y_2 - y_1}{x_2 - x_1} \\
&= \frac{(x + \Delta x)^2 - x^2}{(x + \Delta x) - x} \\
&= \frac{x^2 + 2x\Delta x + \Delta x^2 - x^2}{x + \Delta x - x} \\
&= \frac{2x\Delta x + \Delta x^2}{\Delta x} \\
&= \frac{\Delta x(2x + \Delta x)}{\Delta x} = 2x + \Delta x.
\end{aligned}
$$

Now let's take the limit as Δx approaches 0. As we might have expected, it's an easy plug-in:

$$\lim_{\Delta x \to 0}(2x + \Delta x) = 2x + 0 = 2x.$$

Our result has an x in it. We need to know what x is before we can evaluate it. When we worked out the slope at (1, 1), we were using $x = 1$. Plugging $x = 1$ into $2x$ gives us 2, which confirms the answer we obtained previously.

What do you think the slope is at $x = 2$? Plugging into $m = 2x$ tells us the answer is 4. If we plug in $x = -1$, $2x$ gives us -2. We can now find the slope at every point of the parabola $y = x^2$. Namely, the slope formula is given by $m = 2x$. Check Figure 11-6, where I illustrate several different slopes for $y = x^2$ by sketching in the corresponding tangent lines. As you can see, our generic formula works for both positive and negative values of x.

Figure 11-6.

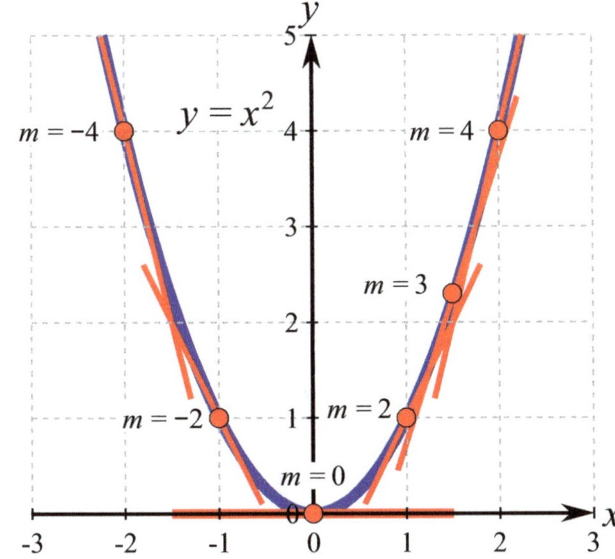

We can now consider working on slightly more challenging functions than $y = x^2$, and that's what we're doing next.

12 A Derivative Work

It's been done, but we're doing it again

In the previous chapter, we looked at the graph of $y = x^2$ and worked out that $m = 2x$ provides a slope formula for the graph. I will use our experience in calculating the slope formula for $y = x^2$ as a model for finding the slope formulas of more general mathematical expressions.

Let's begin by assuming that y can be expressed as some unspecified function of x. You'll recall that this means we can write $y = f(x)$, where $f(x)$ is some function (some formula) that contains x. It could be x^2, as in our earlier example, or it could be more complicated. Or simpler. We just don't need to say what it is at this point. If we graph $y = f(x)$ on an xy coordinate system, we expect to get a curve that moves up and down in various ways, a curve whose steepness (slope!) varies from point to point.

We know that if we want to compute a slope formula for the graph of $y = f(x)$, the trick is to find expressions for Δy and Δx—the vertical and horizontal change, respectively—and compute their ratio. Although the result is just an approximation for the slope at any particular point, we learned that we can take a limit as Δx goes to 0 if we want the exact result. For a given graph, as shown in Figure 12-1, begin by marking off points x and $x + \Delta x$ on the x axis. Then plot the points $(x, f(x))$ and $(x + \Delta x, f(x + \Delta x))$ right on the curve. Write the *difference quotient* (the ratio of differences) that gives us an expression for slope:

$$\frac{\Delta y}{\Delta x} = \frac{f(x + \Delta x) - f(x)}{(x + \Delta x) - x} = \frac{f(x + \Delta x) - f(x)}{\Delta x}.$$

Figure 12-1.

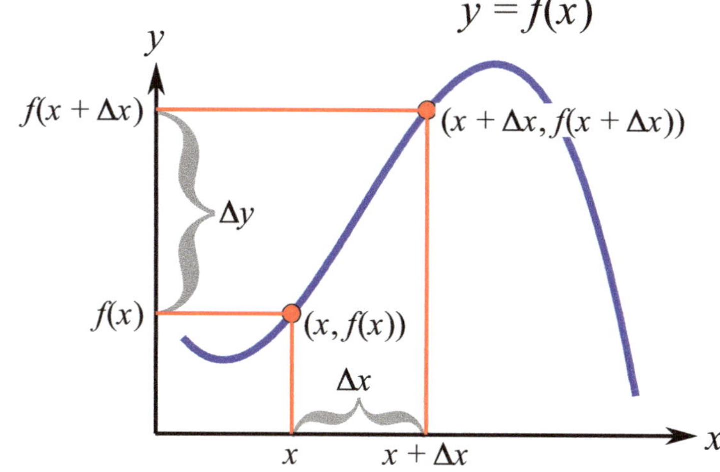

Finally, take the limit as Δx goes to 0. The result is your slope formula for $y = f(x)$:

$$m = \lim_{\Delta x \to 0} \frac{\Delta y}{\Delta x} = \lim_{\Delta x \to 0} \frac{f(x + \Delta x) - f(x)}{\Delta x}.$$

In a nutshell, that's the process that we used in the previous chapter when finding the slope formula for $y = x^2$, but this time we recreated the process without using a specific function. Instead, $f(x)$ served as a generic place holder. As long as we can figure out the limit (and that will depend on the actual choice of $f(x)$), then we will be able to compute a slope formula. Obviously, we need to try out our new tool on some additional examples. First, though, I have to give you a short language lesson.

Talking about derivatives

Mathematicians talk their own language, as I suspect you've already noticed. There is a lot of specialized vocabulary for calculus, some of which we've already seen: words like *integral*, *integrand*, and *limit*. The most important new word that you're going to need right now is *derivative*. That's the word that we use instead of slope formula. Instead of continuing to use m all the time for slope (which I borrowed from the symbols used in algebra because I knew it would be familiar to most readers), I will start using $f'(x)$. The little mark on the f is not an apostrophe. It's a straight little tick mark that mathematicians call a prime symbol. That's why you'll sometimes hear calculus teachers and students talking about "f prime," which is just another way to say "the derivative of f."

Leibniz had his own notation for the derivative (as you might have suspected). He's not responsible for the prime notation. Leibniz preferred to create a symbol more reminiscent of the process by which the derivative was computed. In place of $f'(x)$, he always wrote $\frac{dy}{dx}$. (I know it looks like "dy over dx," and that is what Leibniz wanted you to see, but mathematicians almost always just say "dee y dee x" for short, as if there were no division symbol. It's just more of that mathematical laziness, I suspect.) In Leibniz notation, we could write

$$\frac{dy}{dx} = \lim_{\Delta x \to 0} \frac{\Delta y}{\Delta x},$$

which may give you an idea why Leibniz created it. The d's just replace the Δ's and are supposed to remind you of a quotient, although after the limit has been taken. We'll have other occasions later to use Leibniz notation for derivatives. Most of the time I plan to stick with the more convenient and compact prime notation. In fact, sometimes we'll abbreviate $f'(x)$ as y', which is even simpler.

Derivatives of higher powers

Let's compute the derivative of $y = x^3$. That is, $f(x) = x^3$ and we want to find $f'(x)$. We need to compute $f(x + \Delta x) = (x + \Delta x)^3$. Do you remember learning the binomial theorem in Algebra 2? It's another one of Newton's discoveries. In the case of cubing, Newton's binomial theorem says that $(a + b)^3 = a^3 + 3a^2b + 3ab^2 + b^3$. If we use this pattern with x in place of a and Δx in place of b, we have $(x + \Delta x)^3 = x^3 + 3x^2\Delta x + 3x\Delta x^2 + \Delta x^3$. Let's substitute these results into the formula for the derivative and see if we can get a nice answer:

$$\begin{aligned}
f'(x) &= \lim_{\Delta x \to 0} \frac{\Delta y}{\Delta x} = \lim_{\Delta x \to 0} \frac{f(x + \Delta x) - f(x)}{\Delta x} \\
&= \lim_{\Delta x \to 0} \frac{(x + \Delta x)^3 - x^3}{\Delta x} \\
&= \lim_{\Delta x \to 0} \frac{x^3 + 3x^2\Delta x + 3x\Delta x^2 + \Delta x^3 - x^3}{\Delta x} \\
&= \lim_{\Delta x \to 0} \frac{3x^2\Delta x + 3x\Delta x^2 + \Delta x^3}{\Delta x} \\
&= \lim_{\Delta x \to 0} \frac{\Delta x(3x^2 + 3x\Delta x + \Delta x^2)}{\Delta x} \\
&= \lim_{\Delta x \to 0} (3x^2 + 3x\Delta x + \Delta x^2) \\
&= 3x^2 + 0 + 0 = 3x^2.
\end{aligned}$$

That was a little bit messy, but the results are very clean: If $f(x) = x^3$, then $f'(x) = 3x^2$.

Do we have enough results to see any patterns yet? In the last chapter, we discovered that the derivative of $f(x) = x^2$ is $f'(x) = 2x$. We called it m then because I hadn't introduced the usual derivative notation yet, but this is a good point at which to remember we are still talking about slope. Why? Because I want you to tell me the derivative of $f(x) = x$. Its graph is a straight line, right? In fact, it's the line $y = x$, whose slope is just $m = 1$. (See Figure 12-2.) That is, the derivative of $f(x) = x$ is $f'(x) = 1$. The little table records our results so far:

Figure 12-2.

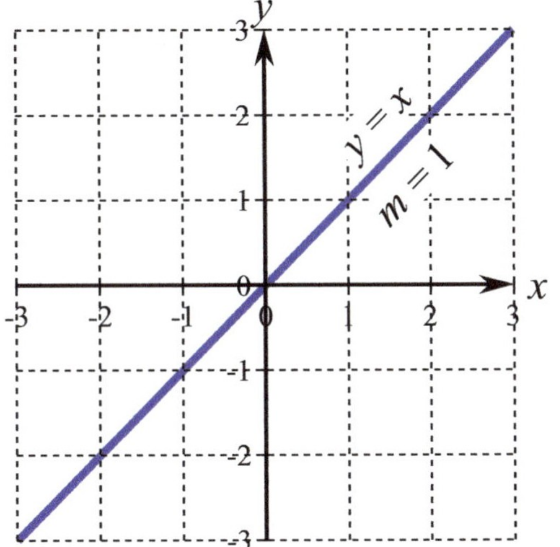

$f(x)$	$f'(x)$
x	1
x^2	$2x$
x^3	$3x^2$

Care to venture a guess for the derivative of $f(x) = x^4$? It's pretty obvious what it should be from the pattern shown in the table. It seems that the original exponent (power) becomes a coefficient and the exponent is dropped by 1. That is, not only should the derivative of $f(x) = x^4$ be $f'(x) = 4x^3$, but the derivative of $f(x) = x^n$ should be $f'(x) = nx^{n-1}$. This is another power rule, but this time it is the *power rule for derivatives*. Our previous power rule, the one for integrals, raised the power by one. This new rule drops it by one. We've just started talking about derivatives and we've already discovered a way in which they are opposites of integrals. Trust me, this will bear watching.

Before we move on to derivatives of other functions (that is, functions other than powers of x), I want to show you two examples of how the power rule works in cases where the power is not a whole number. The first example will showcase a negative power, and the second example will feature a fractional power.

It's not so bad to be negative

The reciprocal function is important enough to have its own calculator key. Every scientific calculator has a reciprocal key. It's the one labeled "$1/x$" or, in some cases, "x^{-1}." Remember that negative powers represent reciprocals. In particular, $\frac{1}{x}$ and x^{-1} mean exactly the same thing. Let's find the derivative of $f(x) = \frac{1}{x}$:

$$\begin{aligned} f'(x) &= \lim_{\Delta x \to 0} \frac{f(x + \Delta x) - f(x)}{\Delta x} \\ &= \lim_{\Delta x \to 0} \frac{\frac{1}{x + \Delta x} - \frac{1}{x}}{\Delta x}. \end{aligned}$$

How do we simplify the quotient so that we can evaluate the derivative? Here it helps to remember the process of using a common denominator to subtract fractions. (See the sidebar.) Let's simplify the difference of fractions in our expression for the derivative:

$$\begin{aligned} \frac{1}{x + \Delta x} - \frac{1}{x} &= \frac{x}{x(x + \Delta x)} - \frac{x + \Delta x}{x(x + \Delta x)} \\ &= \frac{x - (x + \Delta x)}{x(x + \Delta x)} \\ &= \frac{x - x - \Delta x}{x(x + \Delta x)} \\ &= \frac{-\Delta x}{x(x + \Delta x)}. \end{aligned}$$

Now we plug this back into our expression for the derivative and simplify some more. We have to be especially careful in simplifying a fraction containing another fraction, as shown on the next page.

Common denominator

We can't subtract two fractions unless they have the same denominator. Fortunately, we can always alter denominators by multiplying the top and bottom of a fraction by the same number. For example, $\frac{1}{2}$ can be changed into $\frac{5}{10}$ by multiplying both the numerator and the denominator by 5; the form of the fraction is changed, but not its actual value, which remains one-half.

In a more general example, we turn the denominator a and the denominator b into the common denominator ab, after which we can subtract:

$$\frac{1}{a} - \frac{1}{b} = \frac{b}{ab} - \frac{a}{ab} = \frac{b - a}{ab}.$$

$$f'(x) = \lim_{\Delta x \to 0} \frac{\frac{-\Delta x}{x(x+\Delta x)}}{\Delta x}$$

$$= \lim_{\Delta x \to 0} \left(\frac{-\Delta x}{x(x+\Delta x)} \div \Delta x \right) \quad \text{(the top fraction is divided by } \Delta x\text{)}$$

$$= \lim_{\Delta x \to 0} \left(\frac{-\Delta x}{x(x+\Delta x)} \cdot \frac{1}{\Delta x} \right) \quad \text{(dividing by } \Delta x \text{ is the same as multiplying by } \tfrac{1}{\Delta x}\text{)}$$

$$= \lim_{\Delta x \to 0} \frac{-1}{x(x+\Delta x)}. \quad \text{(cancel the } \Delta x\text{'s)}$$

Is the simplified version of the limit ready to evaluate? Can we treat it as a plug-in? Yes:

$$f'(x) = \lim_{\Delta x \to 0} \frac{-1}{x(x+\Delta x)} = \frac{-1}{x(x+0)} = \frac{-1}{x^2}.$$

Therefore the derivative of $f(x) = \frac{1}{x}$ is $f'(x) = \frac{-1}{x^2}$. This is really just another case of the power rule for derivatives. Check it out: If we express $\frac{1}{x}$ in power notation as x^{-1}, then we can see that we are talking about the case where $n = -1$. According to the power rule, the derivative of x^n is given by nx^{n-1}. If we replace n by -1, we have

$$(-1)x^{-1-1} = -1x^{-2} = \frac{-1}{x^2},$$

which is exactly the result we just found. Yes, the power rule for derivatives works for negative powers as well as positive ones.

Of half a mind

I promised we would also look at a fractional power. The one I have in mind is $f(x) = \sqrt{x}$. Since $\sqrt{x}$ can be written as $x^{1/2}$, it really is a fractional power of x. In the previous example, we had to remember how to use a common denominator to subtract two fractions before we could evaluate the limit. This time we will need to simplify a difference of radicals with the *conjugation* technique. (See the sidebar about

The conjugate

Observe what happens when we multiply a difference of radicals by their sum:

$$(\sqrt{a} - \sqrt{b})(\sqrt{a} + \sqrt{b}) = \sqrt{a}\sqrt{a} + \sqrt{a}\sqrt{b} - \sqrt{b}\sqrt{a} - \sqrt{b}\sqrt{b}$$
$$= \sqrt{a}\sqrt{a} - \sqrt{b}\sqrt{b}$$
$$= a - b.$$

As you can see, the simplified product contains no radicals.

The sum and difference of radicals are called *conjugates* of each other because they are a matched set while being opposites—like male/female or husband/wife.

conjugates.) As you'll see, once the radicals in the difference quotient are tamed, we can evaluate the limit and thereby find the derivative of $f(x) = \sqrt{x} = x^{1/2}$:

$$\begin{aligned}
f'(x) &= \lim_{\Delta x \to 0} \frac{f(x + \Delta x) - f(x)}{\Delta x} \\
&= \lim_{\Delta x \to 0} \frac{\sqrt{x + \Delta x} - \sqrt{x}}{\Delta x} \\
&= \lim_{\Delta x \to 0} \frac{\sqrt{x + \Delta x} - \sqrt{x}}{\Delta x} \cdot \frac{\sqrt{x + \Delta x} + \sqrt{x}}{\sqrt{x + \Delta x} + \sqrt{x}} \quad \text{(apply the conjugate)} \\
&= \lim_{\Delta x \to 0} \frac{(x + \Delta x) - x}{\Delta x(\sqrt{x + \Delta x} + \sqrt{x})} \\
&= \lim_{\Delta x \to 0} \frac{\Delta x}{\Delta x(\sqrt{x + \Delta x} + \sqrt{x})} \\
&= \lim_{\Delta x \to 0} \frac{1}{\sqrt{x + \Delta x} + \sqrt{x}}.
\end{aligned}$$

The step where we applied the conjugate to the top and bottom of the quotient resulted in Δx "escaping" from the radicals in the numerator and cancelling the Δx in the denominator. The simplified quotient now has a simple plug-in limit as Δx goes to 0:

$$\begin{aligned}
f'(x) &= \lim_{\Delta x \to 0} \frac{1}{\sqrt{x + \Delta x} + \sqrt{x}} \\
&= \frac{1}{\sqrt{x + 0} + \sqrt{x}} \\
&= \frac{1}{\sqrt{x} + \sqrt{x}} = \frac{1}{2\sqrt{x}}.
\end{aligned}$$

The derivative of $f(x) = \sqrt{x}$ turned out to be $f'(x) = \frac{1}{2\sqrt{x}}$. If we prefer to write these results in fractional-exponent notation, we can say that the derivative of $f(x) = x^{1/2}$ turned out to be $f'(x) = \frac{1}{2}x^{-1/2}$. Why did I suggest this? Because now we see that the derivative of the square root function also conforms to the power rule for derivatives. It's just the case where $n = \frac{1}{2}$ is plugged into the derivative formula nx^{n-1}. (Try it!)

The power rule works for any power of x—positive, negative, whole number, or fraction. If you'll excuse my saying so, the power rule is very powerful. For future reference, here is an expanded version of our summary table:

$f(x)$	$f'(x)$
1	0
x	1
x^2	$2x$
x^3	$3x^2$
$\frac{1}{x}$	$\frac{-1}{x^2}$
$\sqrt{x}$	$\frac{1}{2\sqrt{x}}$
x^n	nx^{n-1}

By the way, you'll notice that I included $f(x) = 1$ and $f'(x) = 0$ in the table. After all, the graph of the function $f(x) = 1$ is just the line $y = 1$, a horizontal line, and any horizontal line has a slope of 0. I chose to use the number 1 for simplicity, but it could just as well have been 2 or 3 or whatever. Whenever $f(x) = c$, where c is some constant, then $f'(x) = 0$, a fact we'll take advantage of in the next chapter.

13 Everything You Always Wanted to Know About Polynomials

Thanks to the power rule, we know that the derivative of $y = x^2$ is given by $y' = 2x$. So what is the derivative of $y = 3x^2$? This second expression is exactly like our first one, except that it includes an extra factor of 3. Does the derivative inherit that 3? I mean, does $y = 3x^2$ have the derivative $y' = 6x$? Let's look at a graph to see if this is a plausible result.

In Figure 13-1, I've drawn in the tangent line with slope $m = 2$ at the point (1, 1) for the curve $y = x^2$. I've also drawn in a line with slope $m = 6$ at the point (1, 3) for the curve $y = 3x^2$. It looks reasonable, right? After all, the curve $y = 3x^2$ rises 3 times as fast as $y = x^2$, so it stands to reason that its derivative (rate of change, or slope) should be 3 times as great.

Figure 13-1.

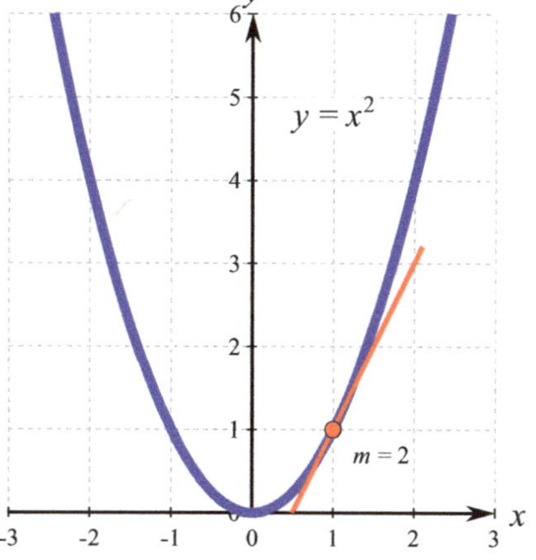

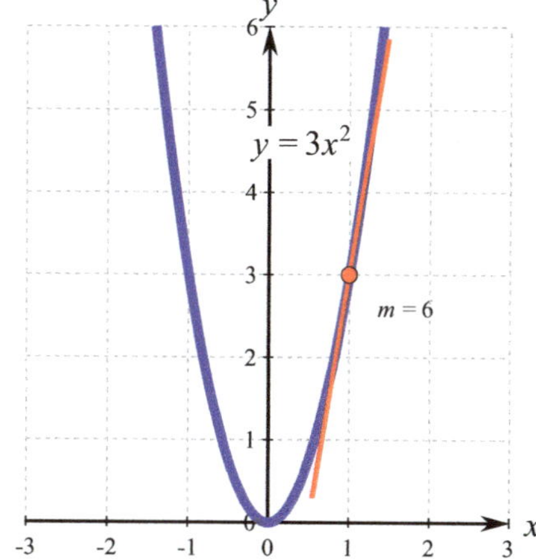

The same thing is true for the equations of lines. We've known how to find the slope of a line since our introductory algebra classes, and we've already taken advantage of that knowledge a few times during our stroll through calculus. Specifically, when a line has an equation written in the form $y = mx + b$, its slope is just m (the coefficient of the x term). Thus we know that the line $y = x$ has slope $m = 1$, the line $y = 2x$ has slope $m = 2$, the line $y = 3x$ has slope $m = 3$, and so on and so forth. We've got the pattern down, and it's part of the general principle that functions with constant factors simply pass those constant factors along to their derivatives. This principle is called the *constant multiple rule for derivatives*, and it can be written down in convenient calculus notation as the equation that relates the derivatives of $cf(x)$ and $f(x)$:

$$(cf(x))' = c \cdot f'(x).$$

Here are some additional examples to make sure we have the idea down:

$$(5x^3)' = 5(x^3)' = 5(3x^2) = 15x^2,$$
$$(4\sqrt{x})' = 4(\sqrt{x})' = 4 \cdot \frac{1}{2\sqrt{x}} = \frac{2}{\sqrt{x}},$$
$$\left(\frac{1}{4}x^4\right)' = \frac{1}{4}(x^4)' = \frac{1}{4} \cdot 4x^3 = x^3.$$

In each of these calculations, the constant that accompanied the power of x was simply multiplied by the derivative obtained by the power rule. The constant multiple rule for derivatives is quite straightforward and might remind you of the constant multiple rule for integrals, in which constants in integrands similarly ended up just multiplying the overall result:

$$\int_a^b cf(x) = c \cdot \int_a^b f(x).$$

In other words, any constant in the integrand could just be factored out.

Adding and subtracting

In algebra we learned that we can combine various powers of x to create mathematical functions called *polynomials*—expressions like $4x + 3$ or $x^3 + x^2$ or $2x^2 - 7x + 5$. How do we take the derivative of a polynomial? Let's see what happens if we try to apply the power rule to each term separately, like this (I'm using the prime notation, which just means "the derivative of"):

$$\begin{aligned}(x^3 + x^2)' &= (x^3)' + (x^2)' \\ &= 3x^2 + 2x.\end{aligned}$$

What do you think? Does that look persuasive to you? If this result is correct, then the derivative of $f(x) = x^3 + x^2$ is given by $f'(x) = 3x^2 + 2x$. We can test this result by plugging in some values of x and examining whether the alleged values of the slope match the graph of the curve. If, for example, we let $x = -1$, we get $f(-1) = (-1)^3 + (-1)^2 = -1 + 1 = 0$ and $f'(-1) = 3(-1)^2 + 2(-1) = 3 - 2 = 1$. Try plugging in $x = 0$ and $x = 1$, too, in which case we get a little table of values, from which I've constructed the graph in Figure 13-2 for you:

x	$f(x)$	$f'(x)$
–1	0	1
0	0	0
1	2	5

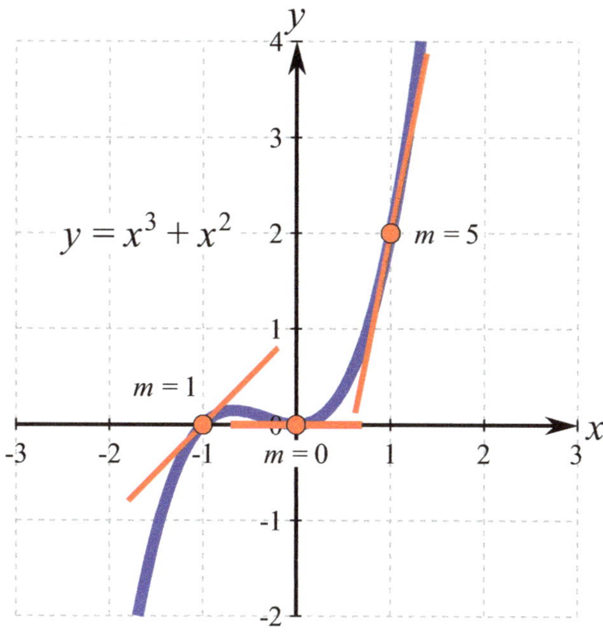

Figure 13-2.

It seems to work, doesn't it? At the point (–1, 0), where we think the slope might be $m = 1$, the sketched-in tangent line matches the curve quite well, as do the other two cases we graphed.

This is just one particular example of the general rule known as the *addition property of derivatives*. We can write it more formally as

$$(f(x) + g(x))' = f'(x) + g'(x).$$

Briefly stated, if we have a function consisting of two simpler functions added together, we can just take the derivative of each part when we want to find the derivative of the whole thing. You could just take my word for it, of course, especially given that I've shown you some examples where it seems to work, but it's good to pin things down more carefully when we can.

It's not difficult to prove the addition property of derivatives. We go back to the limit process for finding derivatives and see what occurs when we try to apply it to $f(x) + g(x)$. Pretend for a moment that we are looking at the graph of $y = f(x) + g(x)$ and are choosing two points on the curve. We pick $(x, f(x) + g(x))$ to be our first point, (x_1, y_1), and $(x + \Delta x, f(x + \Delta x) + g(x + \Delta x))$ to be our second point, (x_2, y_2). Then we find the slope between these two points and take its limit as Δx approaches 0.

Here are the steps:

$$
\begin{aligned}
(f(x)+g(x))' &= \lim_{\Delta x \to 0} \frac{(f(x+\Delta x)+g(x+\Delta x))-(f(x)+g(x))}{\Delta x} \\
&= \lim_{\Delta x \to 0} \frac{f(x+\Delta x)+g(x+\Delta x)-f(x)-g(x)}{\Delta x} \\
&= \lim_{\Delta x \to 0} \frac{f(x+\Delta x)-f(x)+g(x+\Delta x)-g(x)}{\Delta x} \\
&= \lim_{\Delta x \to 0} \frac{f(x+\Delta x)-f(x)}{\Delta x} + \lim_{\Delta x \to 0} \frac{g(x+\Delta x)-g(x)}{\Delta x} \\
&= f'(x)+g'(x),
\end{aligned}
$$

Taking the limit can be "distributed":

$\lim(A+B) = \lim A + \lim B$.

which is exactly the result we were hoping for. We have proved the addition property of derivatives.

There is also a subtraction property of derivatives. The name alone is probably enough to explain to you what it says, but let's write it down for the sake of completeness:

$$(f(x)-g(x))' = f'(x)-g'(x).$$

No surprise there.

By putting together the constant multiple rule with the addition and subtraction rules, we can find the derivative of any polynomial. It doesn't matter how many terms it has, what the powers of x are, or how big the coefficients might be. We can do it, as shown in this example:

$$
\begin{aligned}
(2x^2-7x+5)' &= (2x^2)'-(7x)'+(5)' \\
&= 2(x^2)'-7(x)'+0 \\
&= 2\cdot 2x - 7\cdot 1 \\
&= 4x-7.
\end{aligned}
$$

(Recall that $(5)'=0$ because the derivative of any constant is zero.) Or this example:

$$
\begin{aligned}
(20x^7 + 32x^5)' &= 20(x^7)' + 32(x^5)' \\
&= 20 \cdot 7x^6 + 32 \cdot 5x^4 \\
&= 140x^6 + 160x^4.
\end{aligned}
$$

At this point in our stroll through calculus, we have learned how to find the integral or the derivative of any polynomial. I'm going to list the properties we've used to arrive at this point, both to serve as a handy reference and to underscore the similarities between what we've done so far.

The power rule for derivatives and integrals:

$$
\begin{gathered}
(x^n)' = nx^{n-1}, \\
\int_a^b x^n = \frac{1}{n+1}(b^{n+1} - a^{n+1}).
\end{gathered}
$$

The constant multiple rule for derivatives and integrals:

$$
\begin{gathered}
(cf(x))' = c \cdot f'(x), \\
\int_a^b c \cdot f(x) = c \cdot \int_a^b f(x).
\end{gathered}
$$

The addition property of derivatives and integrals:

$$
\begin{gathered}
(f(x) + g(x))' = f'(x) + g'(x), \\
\int_a^b [f(x) + g(x)] = \int_a^b f(x) + \int_a^b g(x).
\end{gathered}
$$

The subtraction property of derivatives and integrals:

$$
\begin{gathered}
(f(x) - g(x))' = f'(x) - g'(x), \\
\int_a^b [f(x) - g(x)] = \int_a^b f(x) - \int_a^b g(x).
\end{gathered}
$$

With our toolbox loaded with all of these powerful devices, we are ready to tackle the key idea that Newton and Leibniz uncovered at the heart of calculus, a result so important that everyone calls it the *Fundamental Theorem of Calculus*.

14 The Fundamentals of Newton and Leibniz

Isaac and Gottfried find the key

No doubt you've wondered a bit about the curious symmetry between the two power rules. The one for derivatives involves a multiplication by the original exponent and a reduction in the exponent by 1. The power rule for integrals involves an increase in the exponent by 1 and a division by the new exponent. Take another look at them:

$$(x^n)' = nx^{n-1},$$

$$\int_a^b x^n = \frac{1}{n+1}(b^{n+1} - a^{n+1}).$$

Why should a rate of change computation (the derivative) be the opposite of an area calculation (the integral)? What is going on?

Newton and Leibniz both asked these same questions. What's more, they answered them. Our goal in this chapter is to demonstrate why derivatives and integrals have this inverse relationship with each other *and why it is perfectly reasonable*. First, though, let's take the liberty of writing our integral with a variable upper limit of integration, like this:

$$\int_a^x t^n = \frac{1}{n+1}(x^{n+1} - a^{n+1}).$$

As we did back in Chapter 9, we simply plug in x as a substitute for b. (Also as in Chapter 9, we've gone back to using t as our *dummy variable*, because x is in use as the upper limit of integration.) Since the integral now depends on x, a variable, we can call it a function of x. Let's call it $A(x)$, the area function, just as we did in Chapter 9. Let's work with a specific example, say, the case where $a = 1$ and $n = 2$. The region whose area we're computing is shown in Figure 14-1 (where we marked an x value between 2 and 3 for purposes of illustration), and the result is

Figure 14-1.

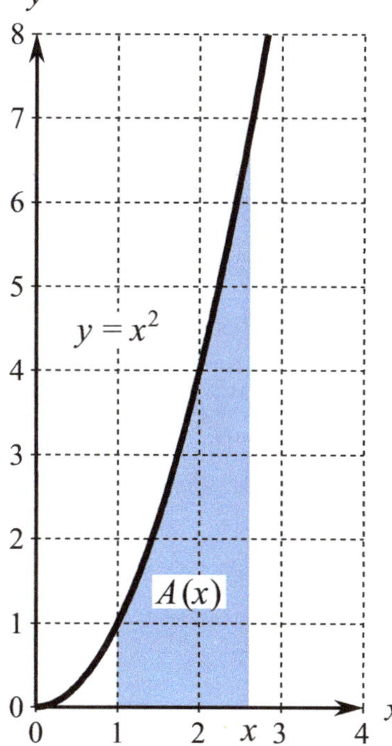

$$A(x) = \int_1^x t^2 = \frac{1}{2+1}(x^{2+1} - 1^{2+1}) = \frac{1}{3}(x^3 - 1).$$

We have now integrated x^2 from 1 to x to obtain the area function $A(x) = \frac{1}{3}(x^3 - 1)$. What do you think the derivative of $A(x)$ might be? It's easy enough to find:

$$\begin{aligned} A'(x) &= \left(\frac{1}{3}(x^3 - 1)\right)' \\ &= \frac{1}{3}(x^3 - 1)' \qquad \left(\text{pull out the factor } \frac{1}{3}\right) \\ &= \frac{1}{3}((x^3)' - (1)') \quad (\text{take the derivative of each term}) \\ &= \frac{1}{3}(3x^2 - 0) = x^2. \end{aligned}$$

We just showed that $A'(x) = x^2$. That is, we just got back the function that we used in the integrand to get $A(x)$ in the first place. In other words,

$$A(x) = \int_1^x A'(t).$$

Of course, we just did it for a very special case, but that's quite a discovery if it works in general. It's saying that the area function is the integral of its own derivative, as though integration cancels out the derivative.

Let's try to demonstrate the inverse relationship of the integral and the derivative in a more general case. In the example with the function x^2, we were able to find the derivative of $A(x)$ quite easily because we had an explicit result to work with. If we try to work out the general case with a generic $f(t)$ in place of t^2, we have to return to basics—namely, we have to use the limit of the difference quotient for $A(x)$. Here's the set-up:

Suppose that a is some nice constant, $f(t)$ is some unspecified function, and $A(x)$ is the area function for the region under the curve $y = f(t)$ for

Figure 14-2.

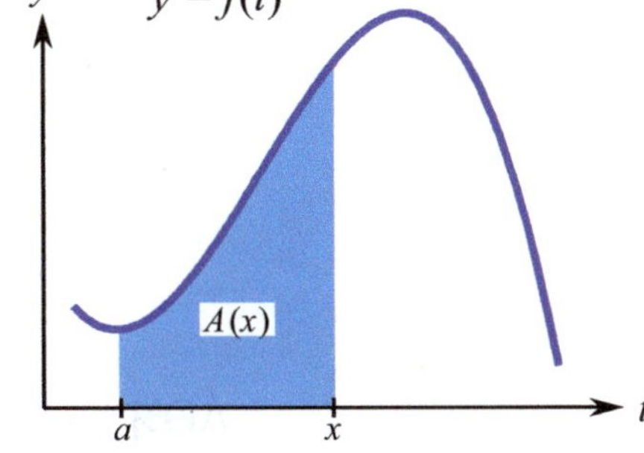

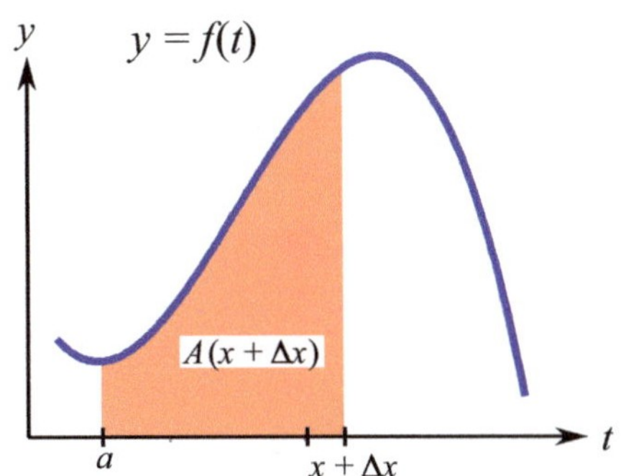

values of t between a and x. In the left part of Figure 14-2, we see what $A(x)$ represents. In the right part, I've illustrated what happens when we replace x with a slightly larger number, $x + \Delta x$.

When we set up the limit of the difference quotient for $A(x)$, we have

$$A'(x) = \lim_{\Delta x \to 0} \frac{A(x+\Delta x) - A(x)}{\Delta x} \quad .$$
$$= \lim_{\Delta x \to 0} \frac{\int_a^{x+\Delta x} f(t) - \int_a^x f(t)}{\Delta x}$$

Now what, exactly, is equal to the difference of the two integrals? It has to be the strip of area in the left part of Figure 14-3. We can call it ΔA. In most cases, we cannot expect it to be a rectangle. However (and this is just a little bit tricky), we should be able to find a number c somewhere between x and $x + \Delta x$, as shown in the right part of Figure 14-3, so that the rectangle with height $f(c)$ and width Δx will exactly equal ΔA. The area of that rectangle is, of course, the height times the width: $f(c)\Delta x$. We can't assume that c is in the exact middle of the interval from x to $x + \Delta x$, but we don't need that. We just need its function value $f(c)$ to give the right height, after which we can write that $\Delta A = f(c)\Delta x$. (There is a mathematical theorem that guarantees the existence of the number c. It's called the *mean value theorem for integrals*.) We substitute this value for ΔA into our limit:

Figure 14-3.

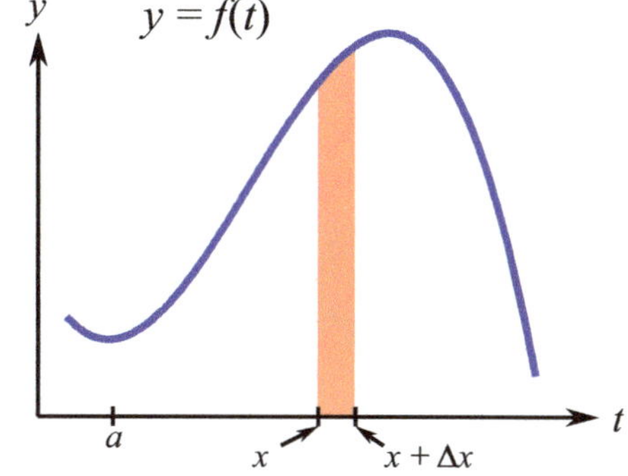

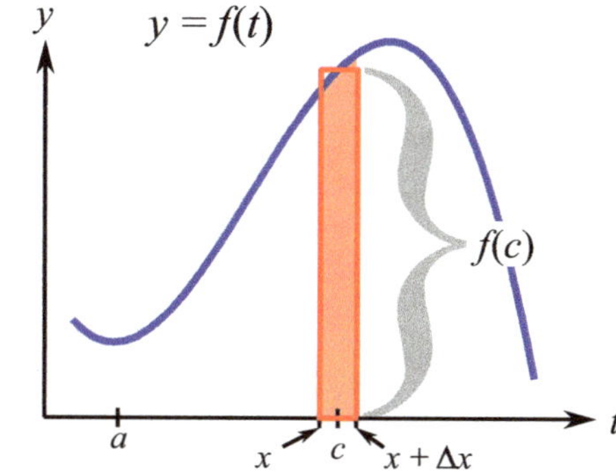

$$A'(x) = \lim_{\Delta x \to 0} \frac{\int_a^{x+\Delta x} f(t) - \int_a^x f(t)}{\Delta x}$$
$$= \lim_{\Delta x \to 0} \frac{\Delta A}{\Delta x}$$
$$= \lim_{\Delta x \to 0} \frac{f(c)\Delta x}{\Delta x}$$
$$= \lim_{\Delta x \to 0} f(c).$$

Well, this is something different, isn't it? The limit is unlike any we've seen before, and it's not exactly a plug-in. In fact, Δx has entirely vanished, and we have nowhere to let it go to zero. Instead, we have to figure out what happens to c. Do you remember where it's supposed to be? I said that we find c between x and $x + \Delta x$. As Δx goes to 0, x and $x + \Delta x$ get closer and closer together, with c trapped in-between. In the limit, therefore, c is stuck between x and $x + 0$, which is no choice at all. In a kind of squeeze play, c has to end up with the same value as x. That allows us to finish our computation:

$$A'(x) = \lim_{\Delta x \to 0} f(c) = f(x).$$

Another way to write this is

$$\left(\int_a^x f(t)\right)' = f(x),$$

which is the general case of the *Fundamental Theorem of Calculus* (often called the FTC for short). The fundamental theorem can be written in more than one way, and this is the first version we proved, so we'll sometimes call it the 1st FTC. Soon we'll see the second version, which should look extremely familiar since it has the same form as the power rule for integrals that we discovered in Chapter 4. Of course, that power rule could be used only with very special functions (powers, of course). That limitation doesn't apply to the fundamental theorems, no matter what version you're talking about.

Although we started by proving the fundamental theorem for a special case (the polynomial function t^2), we made no special assumptions about $f(t)$. Therefore our result is valid provided that we can perform the required computations. Although there are actually some badly behaved functions that would prevent us from computing integrals and limits, I'm pleased to tell you that they are the exception rather than the rule. For the functions we're likely to encounter, the Fundamental Theorem of Calculus will hold up just fine.

The paint-roller theorem

While a mathematical proof (or even the sketch of a proof) is all very nice, it would be even better in our situation to have an intuitive understanding of what the Fundamental Theorem of Calculus is saying. A proof can give us confidence in a result, although it might not always show us just what is going on. As I said at the beginning of the chapter, the FTC is *perfectly reasonable*. One way to see this reasonableness is the *paint-roller theorem*.

If you run a typical paint roller across a flat surface, you get a rectangle of paint. The wider the roller, the wider the rectangle. In Figure 14-4, I've drawn three different cases. The top and bottom paint rollers are perfectly conventional rollers, the top one having a fixed width H and the bottom one having a smaller fixed width h. If you roll each a distance x, you get rectangles of area xH and xh for the top and bottom rollers, respectively.

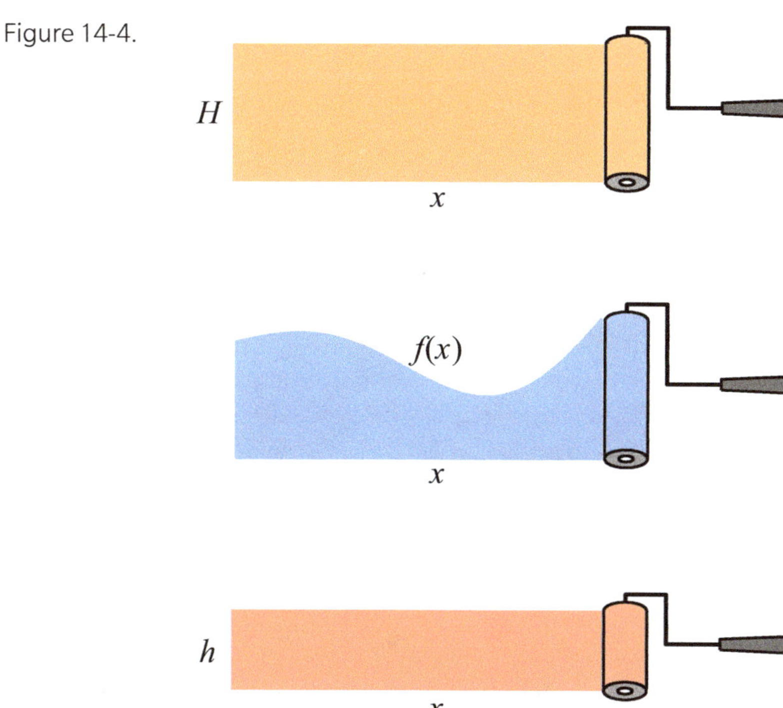

Figure 14-4.

The roller in the middle, however, is a very special case. I suppose we could call it a *magic* paint roller because its width varies as you roll it along. We probably would not have much luck trying to find such a roller at our local hardware store. The middle roller's width is $f(x)$ when it's been rolled x units. It therefore lays out a strip of paint of varying width that corresponds to the area under the curve $y = f(x)$. When $f(x)$ is large, the magic roller leaves a wide strip, while when $f(x)$ is small, the roller leaves a narrow strip. Very interesting.

What is the rate of change of the painted area for each roller? This is a simple question for the top and bottom rollers. For the top roller, the rate of change of the painted area has to be $A'(x) = (xH)' = H$ (remember that H is a constant). For the bottom roller, it's $A'(x) = (xh)' = h$. In other words, the rate of change of the painted area is just the width of the roller (the height of the rectangle). For the magical roller in the middle, the same thing will be true, even though the width of the roller varies. You get more area when the roller is wider, and the rate of change rises and falls in lockstep with the value of $f(x)$; that is, $A'(x) = f(x)$.

The paint-roller theorem is just a different and more intuitive way of stating the Fundamental Theorem of Calculus. We have therefore looked at the FTC from two distinct perspectives—a fairly straightforward mathematical point of view and a thoroughly fictional wall-painting point of view. I hope that together these two aspects of the FTC bring the basic idea into sharp relief for you.

Congratulations! You just got a glimpse of what Newton and Leibniz were privileged to discover and prove over 300 years ago.

15 Down with Derivatives

A new slant on Leibniz notation

I've made no secret of my fondness for Leibniz notation. It's clever and easy to work with. And we have by no means taken advantage of its full power yet. In the wake of our examination of the Fundamental Theorem of Calculus, I should tell you about one more slick feature Leibniz incorporated in his notation.

As we saw in the proof of the FTC, we can evaluate an integral by plugging values into an expression whose derivative is the integrand. Since this is an awkward way to describe things, mathematicians have agreed to call $F(x)$ an *antiderivative* of $f(x)$ if $F'(x) = f(x)$. Because he knew the FTC, Leibniz decided to use similar notation for both the integral and the antiderivative. Following his lead, we can write $\int f(x)$ to mean "antiderivative of $f(x)$." For example, consider this calculus statement:

$$\int x^2 = \frac{1}{3}x^3.$$

When there are no limits of integration (sub- and superscripts on the integral sign), the integral sign just means "antiderivative." The calculus statement we just wrote can, however, be improved. To show this, let me ask you to suppose that $f(x)$ has *two* different antiderivatives, $F(x)$ and $G(x)$. Could you please tell me the derivative of $G(x) - F(x)$? Let's see:

$$(G(x) - F(x))' = G'(x) - F'(x) = f(x) - f(x) = 0.$$

The derivative is zero. What sort of thing has a derivative of zero? Think about it for a second.

That's right. *Constants* have a derivative of zero. Does it go the other way? I mean, if something has a derivative of zero, does it have to be a constant? The good news is that, yes, under the conditions we care about, being a constant and having a derivative of zero are the same thing. We can therefore conclude that $G(x) - F(x) = C$ for some constant C. If we add $F(x)$ to both sides of that equation, we discover that $G(x) = F(x) + C$. What this tells us is that $f(x)$ could have different antiderivatives, but they differ at most by a constant. If I say that the antiderivative of x^2 is $\frac{1}{3}x^3$, you could come back and say that $\frac{1}{3}x^3 + 4$ is just as good an antiderivative—and you'd be right. That's why we routinely write antiderivatives with an extra C term, just in case. It's called the *arbitrary constant of integration,* and it works like this, where I lay out three different antiderivatives, complete with constants:

$$\int x^2 = \frac{1}{3}x^3 + C,$$

$$\int 4x = 2x^2 + C,$$

$$\int \frac{1}{2\sqrt{x}} = \sqrt{x} + C.$$

You can check these by taking the derivative of the results and verifying that each answer is the integrand of the corresponding antiderivative.

Sometimes people like to say that an antiderivative is *a family* of functions, one member of the family for each choice of C. I've drawn several examples from the family of antiderivatives that have the form $y = \frac{1}{3}x^3 + C$, as shown in Figure 15-1.

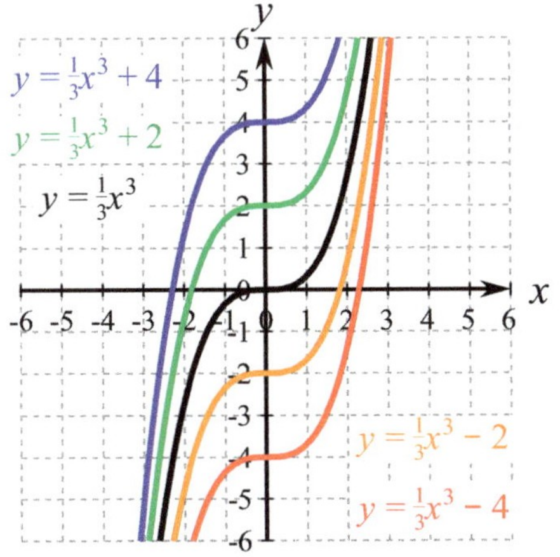

Figure 15-1.

We saw in the previous chapter that the derivative of $\int_a^x f(t)$ is $f(x)$. That is, $\int_a^x f(t)$ is an antiderivative of $f(x)$. This means we can write it as

$$\int_a^x f(t) = F(x) + C.$$

where $F'(x) = f(x)$ and C is an arbitrary constant. If we plug in $x = b$ for the upper limit of integration, we get

$$\int_a^b f(t) = F(b) + C.$$

Since a and b are both constants (even if I didn't tell you exactly what they are), the integral should be a fixed number. That means the C can't be arbitrary anymore, but how can we find out exactly what its value should be? Well, since we plugged in b for x, the upper limit of integration, what occurs if we plug in a instead? This time we get

$$\int_a^a f(t) = F(a) + C.$$

See anything funny about this? The left side isn't the integral from a to b anymore, it's the integral from a to a. If we integrate any function from a to a, we get *nothing*, so the result is zero. (See Figure 15-2.) The equation turns into

$$0 = F(a) + C,$$

which means that $C = -F(a)$. In other words, we have a new way to rewrite the integral from a to b:

$$\int_a^b f(t) = F(b) + C = F(b) - F(a).$$

See Figure 15-3 for a graphical depiction. If you think this new version of the FTC looks familiar, you're right. Consider one of our favorite examples, the one with $f(x) = x^2$ and $F(x) = (\frac{1}{3})x^3$. (It's okay to go back to using x as our dummy variable again because we're no longer using it as a limit of integration.) If we use these, we can see that our new rule for the integral matches what we found earlier:

$$\int_a^b f(x) = F(b) - F(a)$$
$$\int_a^b x^2 = \frac{1}{3}b^3 - \frac{1}{3}a^3.$$

Figure 15-2.

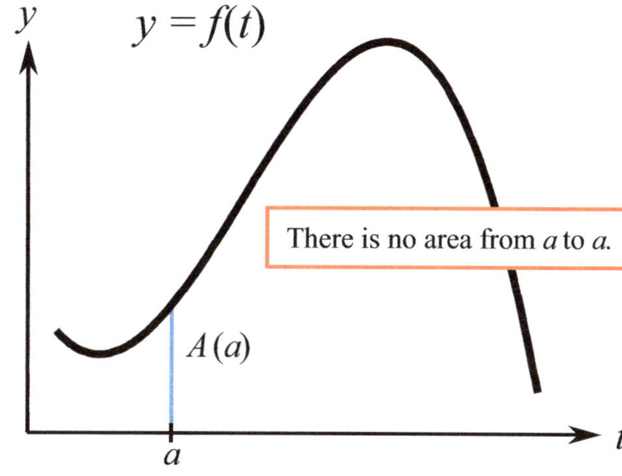

Figure 15-3.

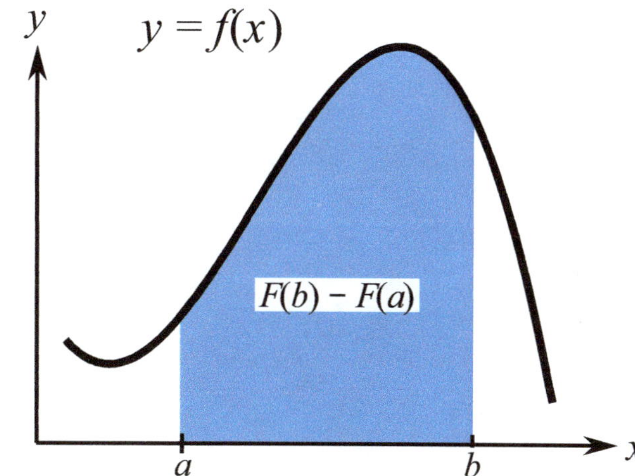

This is one of the first examples of the power rule that we ever saw, now derived as a consequence of the FTC and our new way of writing integrals.

One other traditional piece of math notation is the evaluation bar. It's a vertical bar with either one or two numbers on it, which we place next to a mathematical expression. If there's only one number on the bar, it means we should plug that number into the expression. If there's two numbers on the bar, a subscript and a superscript (much like the integral sign), it means that you should plug in the top number on the bar, plug in the bottom number on the bar, and then subtract the results. This is easier to show with examples than to explain with words:

$$\begin{aligned} x^4\big|_3 &= 3^4 = 81, \\ x^4\big|_1^2 &= 2^4 - 1^4 = 16 - 1 = 15, \\ \frac{1}{3}x^3\Big|_a^b &= \frac{1}{3}b^3 - \frac{1}{3}a^3, \\ \sqrt{x+1}\,\big|_0^3 &= \sqrt{3+1} - \sqrt{0+1} \\ &= \sqrt{4} - \sqrt{1} \\ &= 2 - 1 = 1. \end{aligned}$$

Remembering that $F(x)$ is the traditional notation for an antiderivative of $f(x)$, let's summarize some of the things we've seen so far:

Antiderivative: $\int f(x) = F(x) + C$

FTC (1st version): $\left(\int_a^x f(t)\right)' = f(x)$

FTC (2nd version): $\int_a^b f(x) = F(x)\Big|_a^b = F(b) - F(a)$

Thanks to Leibniz, we can also write the second version of the fundamental theorem as

$$\int_a^b f(x) = \int f(x)\Big|_a^b,$$

which looks almost obvious—two versions of the integral are equal to each other—but it actually required proof. Leibniz chose his notation very deliberately, as a kind of built-in reminder of how calculus works. Pretty clever.

Postscript

Because the integral sign gets used in two distinct ways in calculus, mathematicians usually call $\int_a^b f(x)$ the *definite integral* (where we used to say just "integral") and refer to $\int f(x)$ as the *indefinite integral*. The adjective "definite" in the first case means that we get a specific value or number for our result. The "indefinite" in the second case means that we get a function (or family of functions) instead of just one specific value. The terms "antiderivative" and "indefinite integral" are synonyms that refer to exactly the same thing. I suspect "antiderivative" was coined for the express purpose of referring to a function whose derivative is $f(x)$ without using the word "integral." That way, we math teachers can talk about integrals and antiderivatives without immediately giving away the fact that they are two sides of the same coin. Since Leibniz notation has the FTC built into it, calculus teachers have to work hard to get students to understand it's actually a big deal instead of an obvious fact.

16 Going Around in Circles

A sin isn't always bad

With the Fundamental Theorem of Calculus in hand, we can compute integrals for any function that has an antiderivative. Up to this point, the only functions we really know well are powers of x (and their combinations, as in polynomials). We're stuck with a limited repertory, and it's time to add some new examples to our collection of tricks.

If you have a scientific calculator nearby, take a look at its keyboard. No matter what model you have, you'll see some familiar functions on the keys. (See Figure 16-1.) You can count on finding a key for the square root function, $\sqrt{x}$, and another key for the reciprocal function, $1/x$. In addition, you'll see keys marked "sin," "cos," and "tan." If your math class experience topped out at algebra, then you've never had a formal introduction to the trigonometric functions sine, cosine, and tangent (although "tangent" sounds awfully familiar to someone who's been reading about calculus; there *is* a connection).

In this chapter, we're going to find derivatives and antiderivatives for some of the trigonometric functions. Let's start with the two most popular trig functions, sine and cosine. Figure 16-2 shows the graphs of $y = \sin x$ and $y = \cos x$. As you can see, their graphs are *sine waves*. Big surprise. The sine function passes through the origin, while the cosine function passes through (0, 1) instead. While the sine and cosine curves are identical in shape, they are out of phase (as a mathematician would say).

If we take a very close look at $y = \sin x$, we can quickly identify some important facts about its derivative. Look at Figure 16-3, where I've drawn $y = \sin x$ together with $y = x$. It's rather clear that $y = x$ is tangent

Figure 16-1.

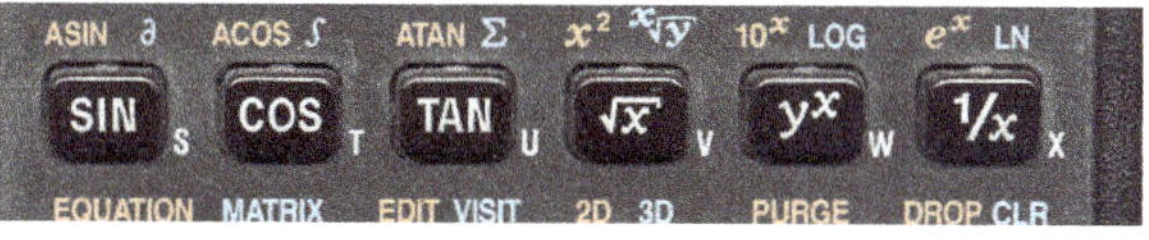

Figure 16-2.

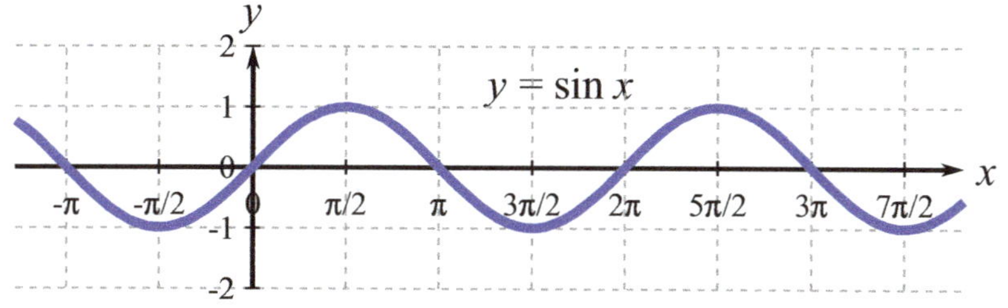

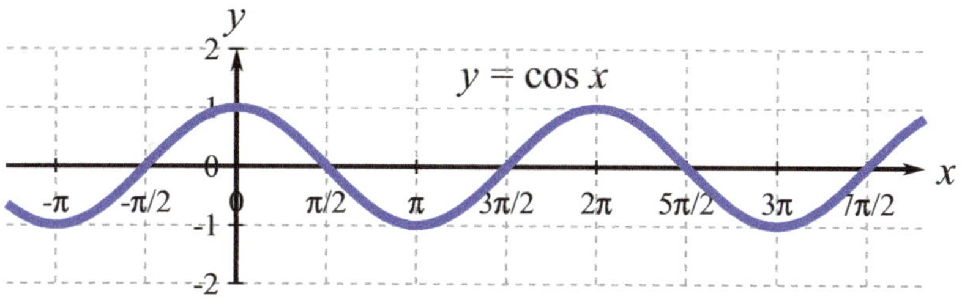

Figure 16-3.

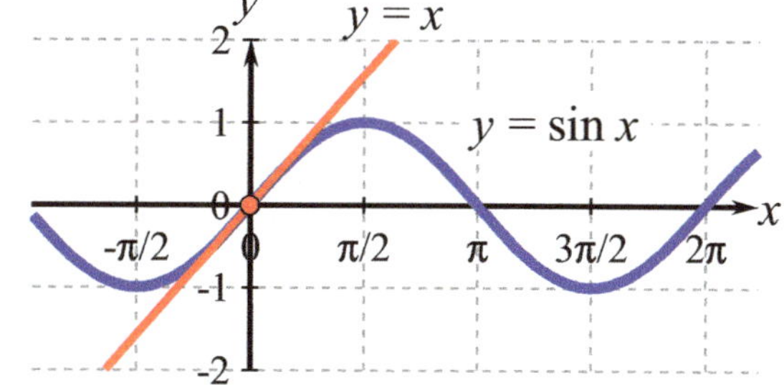

to $y = \sin x$ at the origin. What's more, since we know the slope of $y = x$, we know the derivative (the slope) of $y = \sin x$ at the origin. It's $m = 1$. Although we don't know a formula for $(\sin x)'$ yet, we do know that its value at $x = 0$ has to be 1.

I've redrawn the sine function in Figure 16-4, along with several tangent lines, including the original one at (0, 0). As you can see, the derivative of the sine function must be 0 at $x = -\frac{\pi}{2}, \frac{3\pi}{2}, \frac{5\pi}{2}$, and $\frac{7\pi}{2}$. The derivative must be 1 at $x = 0$ and 2π. The derivative must equal –1 at $-\pi$, π, and 3π. Since the slopes of these tangent lines gave us values of the derivative of sine, I have plotted those values on a second graph in Figure 16-4. That is, each m value in the sine graph corresponds to a dot on the $(\sin x)'$ graph. Do you notice something interesting? The derivative of the sine function looks exactly like the sine curve itself, but only out of phase. If you connect the dots on the graph of $(\sin x)'$ with a smooth curve, you can clearly see that it's the cosine function! I've drawn it in in Figure 16-5.

In brief, we have just seen that $(\sin x)' = \cos x$. A formal demonstration of this fact would have required us to go back to the limit of the difference quotient, with some trigonometric identities needed to figure out the limit's value. Since a formal proof is not what we're going for, I hope you'll be satisfied with the plausibility argument I gave with the graphs of sine and cosine. From now on, we'll assume we know the derivative of sine: You want to know the slope at some point of the sine curve? Then evaluate cosine at that same point and you'll be in business.

Our next step is pretty obvious. Since cosine is the derivative of sine, can we assume that sine is also the derivative of cosine? Not quite. There is a small complication. Let's try the same procedure on the graph of cosine that we did on sine. Certain values of the derivative are easy to read from the cosine graph. We'll plot them on a separate coordinate system and see if we recognize the result. The top graph in Figure 16-6 shows the curve $y = \cos x$, along with several tangent lines at convenient points. We take the values of the slope associated with those tangent

Figure 16-4.

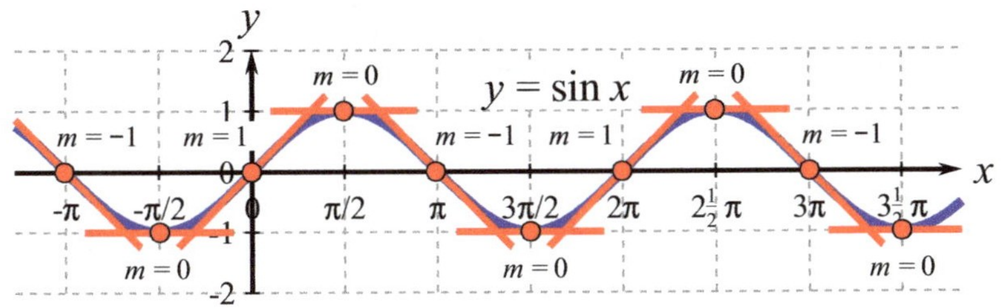

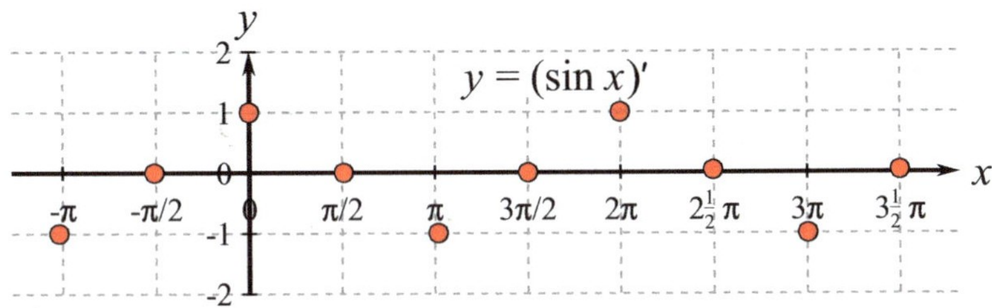

Figure 16-5.

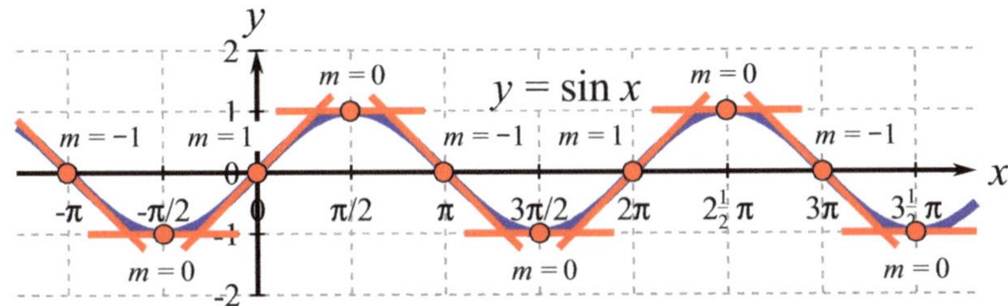

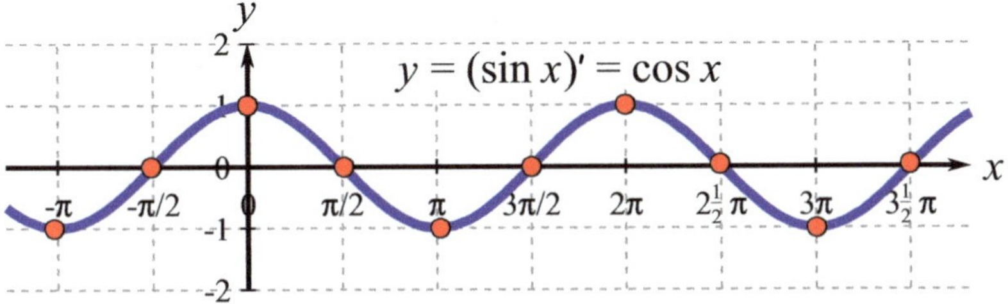

lines and plot them on the bottom graph in Figure 16-6. We can see right away that we are not getting points that lie on the sine curve. Although $(\cos x)'$ passes through the origin, it's not moving in the right direction. The graph of $(\cos x)'$ is not $\sin x$, it's $-\sin x$.

I have filled in the graph of $-\sin x$ in Figure 16-7, and you can see that it matches $(\cos x)'$ perfectly.

Let's write down our two discoveries, the derivatives of sine and cosine:

$$(\sin x)' = \cos x$$
$$(\cos x)' = -\sin x.$$

We now know, however, that any derivative formula is automatically also an antiderivative formula, so let's take note of the fact that we can write a pair of trigonometric indefinite integrals:

$$\int \cos x\, dx = \sin x + C$$
$$\int \sin x\, dx = -\cos x + C.$$

You'll notice that I didn't leave the minus sign in the integrand of the second integral. Like other mathematicians, I prefer to transfer it to the other side of the equation and put it with the result. It would be a good idea if you were to take the derivatives of the result of each antiderivative formula to make sure that you get back the corresponding integrand.

Of course, any time you know antiderivatives, the fundamental theorem says you can compute definite integrals, too. For example, what's the area under one hump of the sine curve? (See Figure 16-8 on the next page.) To find the answer, just integrate $\sin x$ from 0 to π:

$$\int_0^{\pi} \sin x\, dx = -\cos x \Big|_0^{\pi}$$
$$= (\cos \pi) - (-\cos 0)$$
$$= -(-1) - (-1) = 2.$$

Figure 16-6.

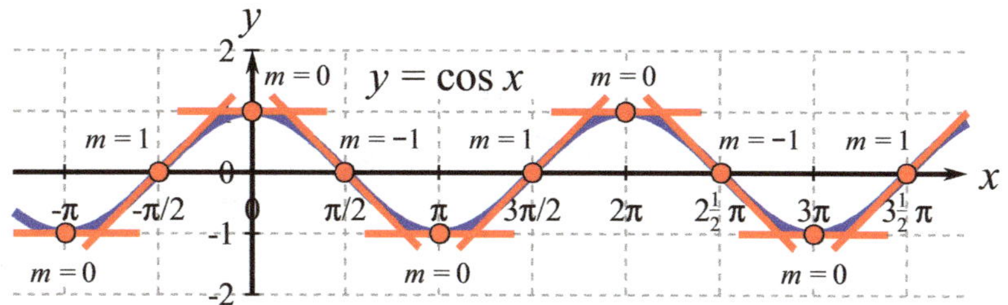

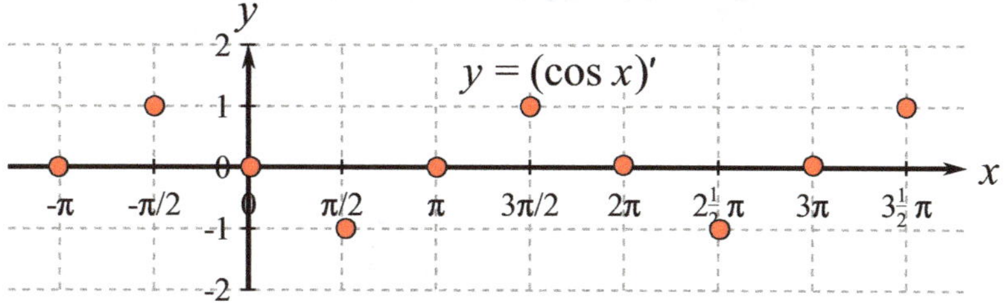

Figure 16-7.

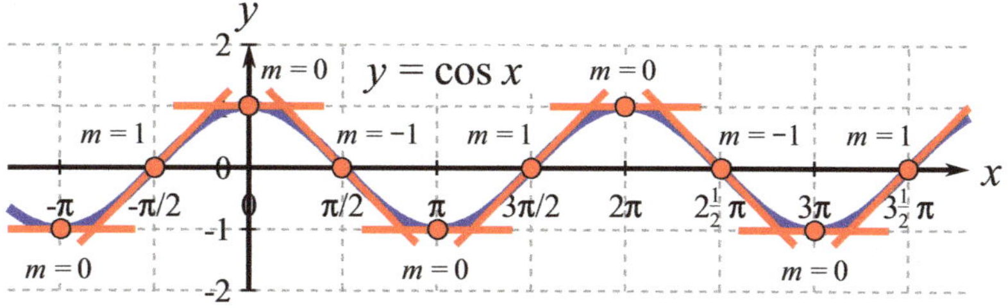

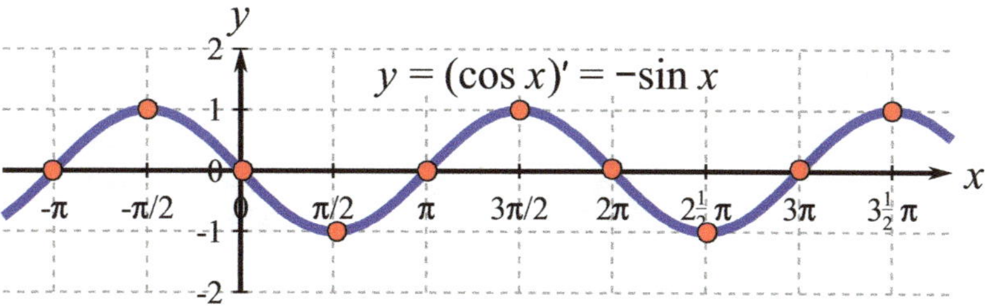

It would be nice at this point if we could charge ahead and work out the derivative of the tangent function, tan x, which is the ratio of sine and cosine. However, there are a few important facts from trigonometry that I can't assume you have at your fingertips. In addition, there is a special rule for computing the derivatives of ratios (the *quotient* rule) that we don't have in our calculus toolbox yet.

Figure 16-8.

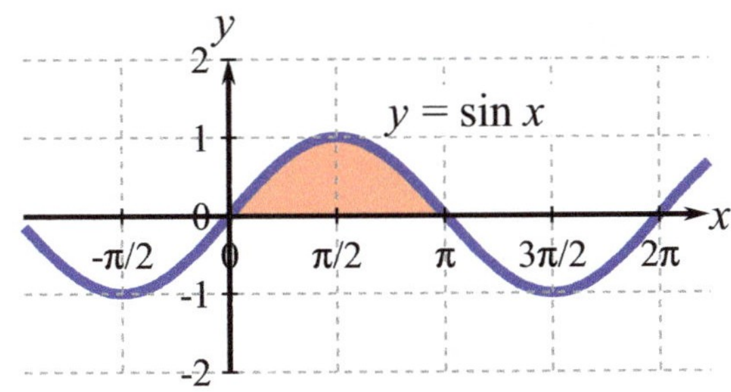

To prepare us for the next stage of our stroll through calculus, I therefore need to take care of both of these problems. I'll finish this chapter off with a brief review of some facts from trigonometry, and in the next chapter, I'll tell you about the product and quotient rules for derivatives. Then we'll be able to compute the derivative of the tangent function (and a whole lot more besides).

Some basic facts from trigonometry

We looked at several graphs of sine and cosine functions earlier in this chapter. Did you wonder why the x axes were all labeled in terms of multiples of π? There's a good reason for that. While most of us first encounter angles measured in degrees when it comes to geometry and trigonometry, calculus insists on using *radian measure*. Let me describe radian measure to you, explain its use in defining trig functions like sine and cosine, and then show why radians are much, much better than degrees when it comes to calculus.

Figure 16-9.

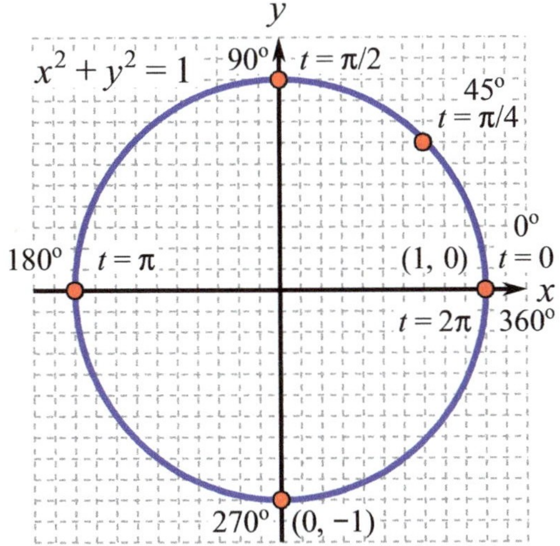

We begin with the unit circle—a circle of radius 1—centered at the origin of an xy coordinate system, as shown in Figure 16-9. In Algebra 2 we learned that its equation is $x^2 + y^2 = 1$.

What is its circumference? Since the circumference of a circle is given by $C = 2\pi r$, where r is the circle's radius, we can plug in $r = 1$ to discover that $C = 2\pi$ for the unit circle. Therefore, if we were to trace around the perimeter of the circle, beginning at (1, 0), moving counterclockwise, and continuing until we return to our starting point, we will travel a distance of 2π units. Since we already know that it takes

360° to go all the way around a circle, we can say that the distance of 2π corresponds to 360°.

How shall we keep track of our trek around the circle? Since x and y are already busy serving as coordinates, we'll use t, the counterclockwise distance around the circumference. As marked on the circle in Figure 16-9, $t = 0$ corresponds to both 0° and the point (1, 0), $t = \frac{\pi}{4}$ corresponds to 45°, $t = \frac{\pi}{2}$ to 90°, $t = \pi$ to 180°, etcetera, until $t = 2\pi$ brings us full circle (literally) back to our starting point of (1, 0), where 360° coincides with 0°. The values of t are what we call *radian measure*. Unlike degrees, radians actually measure the distance traveled around the circumference of a unit circle; but, like degrees, radians can also be used to indicate the magnitude of an angle. Radians thus serve double duty, giving us a way to measure angles while also tipping us off to unit circle distance. (Mathematicians have a very strict rule about angle measure: If you write an angle as just a number without any units, like 10, that means 10 *radians*. If you mean degrees, you must say so: 10°. Don't omit the degree symbol in that case.)

Now for the sine part. And cosine, too. Pick a value of t and go to the corresponding point on the unit circle. For example, if we choose $t = \frac{\pi}{2}$, the corresponding point is (0, 1), as shown in Figure 16-10. The x coordinate of the point is the cosine of t and the y coordinate is the sine of t. For $t = \frac{\pi}{2}$, therefore, the cosine is 0 and the sine is 1. We normally write these as $\cos(\frac{\pi}{2}) = 0$ and $\sin(\frac{\pi}{2}) = 1$. You can also see from the figure that $\cos(0) = 1$, $\sin(0) = 0$, $\cos(\pi) = -1$, $\sin(\pi) = 0$, $\cos(\frac{3\pi}{2}) = 0$, and $\sin(\frac{3\pi}{2}) = -1$.

I've also included the special case $t = \frac{\pi}{4}$, which corresponds to 45°. According to the figure, we have $\cos(\frac{\pi}{4}) = \sin(\frac{\pi}{4}) =$. You might

Figure 16-10.

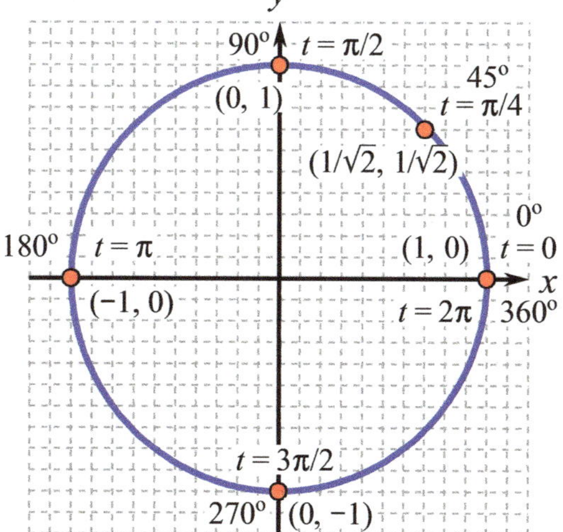

remember this value. We know that $\cos(\frac{\pi}{4})$ and $\sin(\frac{\pi}{4})$ have to be equal to each other because the point on the unit circle corresponding to $t = \frac{\pi}{4}$ (also known as 45°) lies on the line $y = x$. Let's call this unknown common value a, so we're looking at the point (a, a) on the circle $x^2 + y^2 = 1$. If we plug in the point (a, a), we get

$$\begin{aligned} a^2 + a^2 &= 1 \\ 2a^2 &= 1 \\ a^2 &= \frac{1}{2} \\ a &= \pm\frac{1}{\sqrt{2}}. \end{aligned}$$

We can ignore the negative possibility because our graph shows that a has to be positive.

The unit-circle definition of sine and cosine has an immediate consequence. Since each point on the unit circle must satisfy the circle's equation, we always know that $x^2 + y^2 = 1$. But the x value gives us $\cos t$ and the y value gives us $\sin t$, so that means that

$$(\cos t)^2 + (\sin t)^2 = 1.$$

This is what mathematicians call a trigonometric identity. A trig identity is simply an equation involving trig functions that is always true, no matter what angle t (or whatever other variable you're using) is plugged into it.

Once the sine and cosine have been defined, we can move on to the tangent function, which is their ratio:

$$\tan t = \frac{\sin t}{\cos t}$$

If we look at the unit circle again, it's not too difficult to see that most choices of an angle t must correspond to a right triangle whose vertices are (0, 0), ($\cos t$, 0), and ($\cos t$, $\sin t$), as shown in Figure 16-11. (I marked the angle t in its traditional location near the origin, highlighting it with a short arc, but in reality t is the length measured along the circumference of the circle.) When we look at the definition of the tangent function again, we see that it equals rise over run. That is, the tangent function gives the slope of the hypotenuse (the long side, whose length equals 1 because of the unit circle).

Figure 16-11.

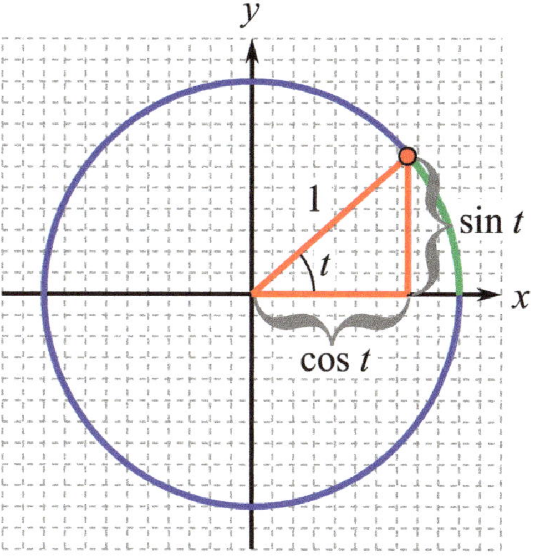

I said that *most* choices of an angle t correspond to a right triangle. One exception is the value $t = \frac{\pi}{2}$, which takes us to the point (0, 1) on the y axis. (Try to draw the triangle on the graph. You can't!) Then, as we've already seen, $\cos t = \cos(\frac{\pi}{2}) = 0$ and $\sin t = \sin(\frac{\pi}{2}) = 1$. If we try to compute $\tan(\frac{\pi}{2})$, we get

$$\tan \frac{\pi}{2} = \frac{\sin \frac{\pi}{2}}{\cos \frac{\pi}{2}} = \frac{1}{0},$$

which is undefined. Since the radius from (0, 0) to ($\cos t$, $\sin t$) is vertical in this case, and vertical lines have undefined slope (they have zero run, and you can't divide by zero), the result for the value of the tangent function at $\frac{\pi}{2}$ is consistent with the earlier rise-over-run definition of slope.

On the next page, without further fuss, is a table of some of our most popular trig function values (I give the t values in three forms, first as an exact fraction of π, second as a decimal approximation, and third as the equivalent degree measure).

t (exact)	*t* (approx.)	*t* (degrees)	*cos t*	*sin t*	*tan t*
0	0	0°	1	0	0
$\frac{\pi}{6}$	0.5236	30°	$\frac{\sqrt{3}}{2}$	$\frac{1}{2}$	$\frac{1}{\sqrt{3}}$
$\frac{\pi}{4}$	0.7854	45°	$\frac{1}{\sqrt{2}}$	$\frac{1}{\sqrt{2}}$	1
$\frac{\pi}{3}$	1.0472	60°	$\frac{1}{2}$	$\frac{\sqrt{3}}{2}$	$\sqrt{3}$
$\frac{\pi}{2}$	1.5708	90°	0	1	undef.
π	3.1416	180°	–1	0	0
$\frac{3\pi}{2}$	4.7124	270°	0	–1	undef.
2π	6.2832	360°	1	0	0

Trigonometric functions are *periodic*. That is, they repeat their values over and over again. In the case of sine and cosine, they repeat every 2π radians (360°), while the tangent repeats every π radians (180°). One of the most important reasons for using radians in calculus is the matter of scale. It takes 360 units before sine begins to repeat if we use degree measure, but sine begins to repeat in only about 6.28 units when radian measure is used. Since the sine can never be greater than 1 or less than –1, degree measure is grossly disproportionate to the range of values of the sine (or cosine, for that matter).

Check out what I mean in Figure 16-12, where I've drawn the graph of $y = \sin x$ in degree mode. I had to grossly exaggerate the scale of the y axis so that you can even see it relative to the huge span of the x axis. Indeed, where the value of the sine reaches its maximum of 1 unit (at 90°) should be only $\frac{1}{360}$ as high as the width of the x axis from 0° to 360°.

Figure 16-12.

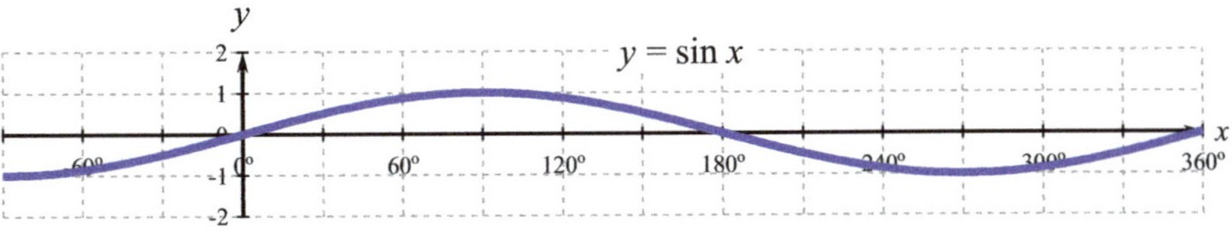

Although most of us are used to measuring angles with degrees, we need to recognize that degree measure is artificial. Perhaps 360 degrees for a full circle is convenient in some ways because it can be so easily divided up (360 has lots of divisors), but it's a largely arbitrary choice. If we picked 400 instead, for example, we could have the measuring system known as gradians or grads. (Yes, there are 400 grads in a full circle; grad measure is included in a number of scientific calculators—most Hewlett-Packards, for example—but it's not very popular.)

Is there some sense in which radian measure is not as arbitrary as degree measure? After all, 2π seems an awkward number for a full circle. In fact, the choice of radian measure is crucial in calculus. Our discussion of the derivatives of sine and cosine was entirely dependent on the observation that $y = \sin x$ has $y = x$ as its tangent line at the origin. (Look back at Figure 16-3 again.) The graphs were virtually indistinguishable near (0, 0). This would not have happened with any other angle measure. To drive the point home, look at this table of values of $\sin x$ when x is a small number:

x	0.1	0.01	0.001	0.0001
$\sin x$	0.099833	0.0099998	0.0009999998	0.000099999999

Provided that we use radian measure, $\sin x$ and x are practically equal for small values of x. This useful fact was our starting point for the calculus of trig functions. It finds applications in other fields as well (such as in the formula from physics for the period of a swinging pendulum, where $\sin x$ is assumed to equal x as long as the pendulum doesn't swing too wide).

Radian measure turned out to be an extremely natural way to work with angles and trigonometry. As one math professor explained it to me, even *Star Trek*'s Mr. Spock would have learned about radian measure as a young student on Vulcan. It's intrinsic to the circle and the natural angle measure for calculus applications. I can't tell you the Vulcan word for “radian,” but whatever Spock called it, it’s built into every circle in the universe.

Since degree measure is arbitrary (and in that sense unnatural), we avoid it in calculus. Radian measure is what works.

17 Go Forth and Multiply

Divide and conquer, too

After a mathematical feast like the previous chapter, it will be nice to ease up somewhat and treat some gentler topics. I am going to show you the product and quotient rules for derivatives. These are extremely important rules (and the quotient rule will let us find the derivative of tan x), but they are not particularly difficult to understand, even if the product rule initially threw the eminent Leibniz for a loss.

Suppose you have a function $h(x)$ that is the product of two other functions, $f(x)$ and $g(x)$. How can we find the derivative of $h(x)$? That is, how do we figure out the formula for

$$h'(x) = (f(x)g(x))'?$$

I propose to work out the formula by going all the way back to our friend the rectangle, who played such a key role in getting us started at the beginning of this book. We remember, of course, that the area A of a rectangle is the product of its length L and its width W:

$$A = LW.$$

Suppose that L is constant, but W is changing. What, then, is the change in A? Well, we already know the constant multiple rule for derivatives, the one that says $(cf(x))' = cf'(x)$. (That's the rule that says constant factors just go along for the ride, completely unaffected by taking the derivative.) Since we're assuming that L is a constant factor, we can write an equation for the rate of change (the derivative) of A:

$$A' = (LW)' = LW'.$$

That is, the rate of change of A is just L times the rate of change of W. Let's look at an actual example:

Consider a rectangle of length $L = 4$ ft, and suppose that the width is $W = 3$ ft but is changing at a rate of $W' = 0.25$ ft/sec. Look at the "time lapse" illustration of the rectangle's size in Figure 17-1, where I show the situation at $t = 0$, 1, and 2 seconds. We easily see that the area's rate of change is 1 ft²/sec. If we check our formula, plugging in $L = 4$ ft and $W' = 0.25$ ft/sec, we get

$$\begin{aligned} A' &= L \cdot W' \\ &= (4\text{ ft})(0.25\text{ ft/ sec}) \\ &= 1.00\text{ ft}^2/\text{ sec}. \end{aligned}$$

It works. The rate of change of area equals the product of the constant length and the rate of change of the width.

It's not too hard to see that we would get something similar if the roles were reversed—if L were changing and W were constant: $A' = L'W$. So what would happen if *both* L and W are changing? What would be the total rate of change of A? At one point, Leibniz thought the answer would be $A' = L'W'$, the product of the derivatives, but then he realized: No, the total rate of change would be the *sum* of the change caused by the length and the change caused by the width. Add them up! And that's how the *product rule for derivatives* (also known as the *Leibniz rule*) was derived:

$$A' = (LW)' = L \cdot W' + L' \cdot W.$$

The product rule is more often stated in function notation, as follows: If $h(x) = f(x)g(x)$, the derivative of $h(x)$ is given by

$$h'(x) = (f(x)g(x))' = f(x)g'(x) + f'(x)g(x).$$

Let's test our new derivative rule on a suitable function, like the product of x and the sine of x. If $y = x \sin x$, then the product rule says that

Figure 17-1.

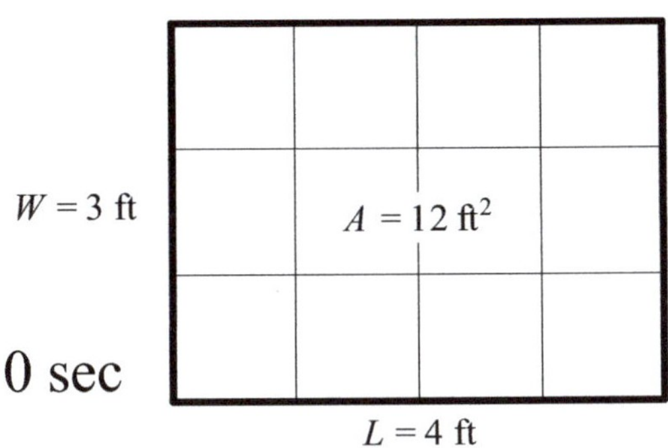

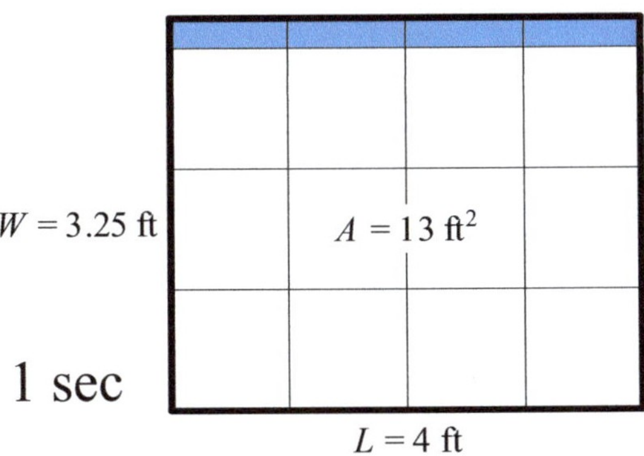

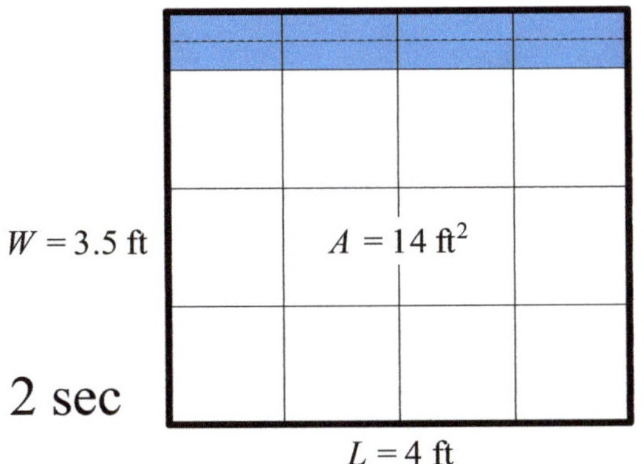

$$
\begin{aligned}
y' &= (x \sin x)' \\
&= x(\sin x)' + x' \sin x \\
&= x \cos x + 1 \cdot \sin x \\
&= x \cos x + \sin x.
\end{aligned}
$$

The derivative came out a bit more complicated than the original function, which is a little different from what we've been used to with polynomials. (When we take the derivative of a polynomial, the polynomial always drops a degree and gets simpler.) If we check whether our result makes sense for a particular point, try plugging $x = \frac{\pi}{2}$ into the result for y′:

$$
\begin{aligned}
y'\big|_{\pi/2} &= (x \cos x + \sin x)\big|_{\pi/2} \\
&= \frac{\pi}{2} \cos \frac{\pi}{2} + \sin \frac{\pi}{2} \\
&= \frac{\pi}{2} \cdot 0 + 1 = 1.
\end{aligned}
$$

According to the formula for y', the slope of $y = x \sin x$ when $x = \frac{\pi}{2}$ is $m = 1$. If you look at Figure 17-2, you'll see that it makes good sense.

We can check out other points as well. If we evaluate y' at $x = \pi$, we get a slope of $\pi \cos \pi + \sin \pi = -\pi$. That's a fairly steep negative slope. As you can see, I've included the tangent line at $(\pi, 0)$ in Figure 17-2. Its slope is certainly negative and fairly steep.

Figure 17-2.

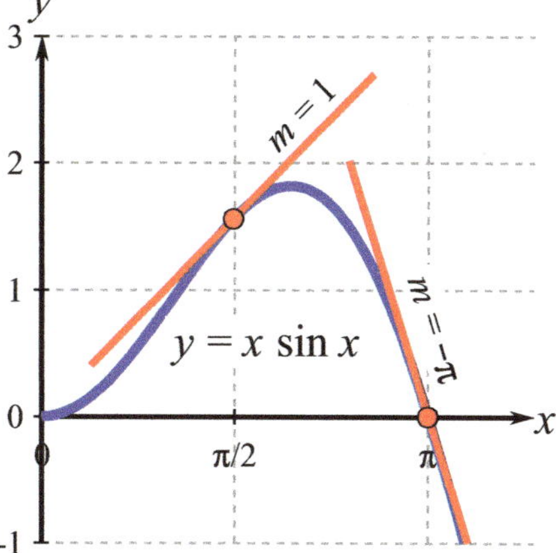

Newton vs. Einstein?

If you've ever taken a natural science or physics class, you may have been introduced to $F = ma$ as Newton's second law of motion. The equation says that force F is equal to the product of mass m and acceleration a. (To avoid including a constant of proportionality, you have to use the right units for force, mass, and acceleration. The SI units commonly referred to as the "metric system" fit the bill.) Many people understand that Einstein's theory of relativity overthrew Newtonian physics and exposed the flaws in his laws. That conclusion is not wholly justified.

The usual reason for picking on Newton's second law of motion is Einstein's discovery that mass is affected by velocity: The faster you go, the more mass increases, until near the speed of light it wants to become infinite (which is why you can't go as fast as the speed of light). However, whether or not Newton believed that mass was constant, he didn't really assume that in his own formulation of the second law. He stated it in greater generality than the $F = ma$ version that is common today.

Petr Beckmann, the irascible author of *A History of Pi*, makes the point explicitly:

> **Contrary to widespread belief, Newton's laws of motion are not contradicted by Einstein's Theory of Special Relativity. Newton never made the statement that force equals mass times acceleration. His Second Law says**
>
> $$F = d(mv)/dt$$
>
> **and Newton was far too cautious a man to take the *m* out of the bracket.**

Beckmann is explaining that Newton actually said that force is equal to the rate of change of mv, the quantity known as momentum. If m is constant, then we can simplify the second law into its usual contemporary form:

$$\begin{aligned} F &= \frac{d}{dt}(mv) \\ &= m\frac{d}{dt}(v) \\ &= m\frac{dv}{dt} = ma. \end{aligned}$$

If, however, we do not assume m to be constant, Newton's second law requires the application of the product rule:

$$\begin{aligned} F &= \frac{d}{dt}(mv) \\ &= m \cdot \frac{d}{dt}(v) + \frac{d}{dt}(m) \cdot v \\ &= m \cdot \frac{dv}{dt} + \frac{dm}{dt} \cdot v \\ &= ma + \frac{dm}{dt} \cdot v. \end{aligned}$$

Thus we have a second term in Newton's second law, one which takes into account the possible change in mass by including $\frac{dm}{dt}$.

Did Newton anticipate that something like this might happen? Beckmann thinks Sir Isaac was being careful and avoiding making too many assumptions, but it seems unlikely to me that Newton was deliberately leaving room for the possibility that mass might change under the influence of motion. Of course, I do not know. What do you think? Was Newton saving some space for Einstein to come along and fill more than 200 years later?

The quotient rule

If there's a product rule, then it seems there should also be a quotient rule. There is, and it's an immediate consequence of the rule for products. We begin by considering the case where a function $h(x)$ is defined to be the quotient of two other functions, $f(x)$ and $g(x)$. We know that

$$h(x) = \frac{f(x)}{g(x)} \quad \text{implies} \quad h(x)g(x) = f(x).$$

Let's apply the product rule to $h(x)g(x) = f(x)$ and see if we can discover a rule for $h'(x)$ on the next page. It will take some algebra to solve for $h'(x)$, but the quotient rule will quickly fall out.

$$(h(x)g(x))' = (f(x))'$$
$$h(x)g'(x) + h'(x)g(x) = f'(x) \quad \text{(apply the product rule)}$$
$$h'(x)g(x) = f'(x) - h(x)g'(x) \quad \text{(now solve for } h'(x)\text{)}$$
$$h'(x) = \frac{f'(x) - h(x)g'(x)}{g(x)}$$
$$h'(x) = \frac{f'(x) - \frac{f(x)}{g(x)}g'(x)}{g(x)} \quad \text{(remember that } h(x) = f(x)/g(x)\text{)}$$
$$h'(x) = \left(\frac{f(x)}{g(x)}\right)' = \frac{g(x)f'(x) - f(x)g'(x)}{(g(x))^2}.$$

And that, ladies and gentlemen, boys and girls, is the quotient rule. Can you tell that it's not very popular with calculus students? The quotient rule is a lot more complicated than the product rule, but too important to neglect. Let's use it right now to find the long-awaited derivative of tan x. We replace every occurrence of $f(x)$ in the formula with sin x and every occurrence of $g(x)$ with cos x:

$$(\tan x)' = \left(\frac{\sin x}{\cos x}\right)' = \frac{(\cos x)(\sin x)' - (\sin x)(\cos x)'}{(\cos x)^2}$$
$$= \frac{(\cos x)(\cos x) - (\sin x)(-\sin x)}{(\cos x)^2}$$
$$= \frac{(\cos x)^2 + (\sin x)^2}{(\cos x)^2} \quad \text{(remember that } (\cos x)^2 + (\sin x)^2 = 1\text{)}$$
$$= \frac{1}{(\cos x)^2}.$$

If you remember that the reciprocal of the cosine function is called the secant function in trigonometry, you could also write $(\tan x)' = (\sec x)^2$, which is neater to write but has the disadvantage of using a function that doesn't appear on your calculator keyboard. Notice that the derivative of tan x is always positive according to this result, because a square can never be negative. Since I haven't shown you the graph of $y = \tan x$ up to this point, it's time to present it to you for consideration. As you can see in Figure 17-3, the graph of $y = \tan x$ has some breaks in it (the places

Figure 17-3.

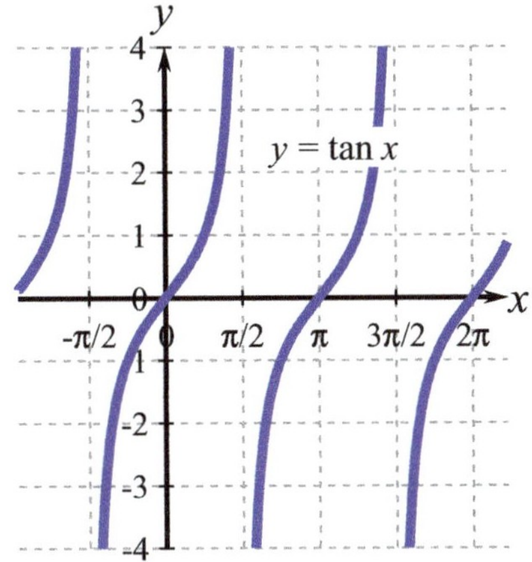

where it doesn't exist, as we saw earlier), but we observe that its graph always has positive slope. Our result for the derivative of the tangent function appears to be consistent with the behavior of the graph.

Postscript

We treated the three most popular trig functions, sine, cosine, and tangent. There are three others. We mentioned secant, which is the reciprocal of cosine. The other two are cosecant, which is the reciprocal of sine ($\csc x = \frac{1}{\sin x}$), and cotangent, which is the reciprocal of tangent ($\cot x = \frac{1}{\tan x}$). While secant is useful because of its relation to tangent, the other two so-called "reciprocal" functions are mostly neglected. Still, now that you know the quotient rule, you could easily find the derivatives of all of them. If you want to test your knowledge, here is a summary of the six trig functions and their derivatives. Try to find the derivatives we didn't cover.

$f(x)$	$f'(x)$
$\sin x$	$\cos x$
$\cos x$	$-\sin x$
$\tan x$	$(\sec x)^2$
$\sec x$	$\sec x \tan x$
$\csc x$	$-\csc x \cot x$
$\cot x$	$-(\csc x)^2$

18 Only as Strong as the Weakest Link

Hooking functions together with the chain rule

A girl is pulling a wagon containing her little brother. He sits amid a pile of pebbles, which he is picking up and dropping on the ground as his sister pulls him along. The boy drops the pebbles at such a rate that he places 4 pebbles on every foot of ground traveled. If the girl is strolling along at a rate of 3 ft/sec, how fast is her brother dropping pebbles?

Okay, this may not be the most practical or realistic of problems, but it's simple to understand and easy enough to work out with simple arithmetic. Soon I'll rewrite it in calculus notation and this little problem will open the gateway to our final and most important derivative rule.

Let's write down what we have so far: The girl's rate of travel is 3 ft/sec, the pebbles are strewn with a density of 4 pebbles/ft (sorry, I don't know an official abbreviation for "pebbles" as a unit of measurement), and we want to figure out the boy's rate of pebble-dropping in units of ... what? Since I asked how "fast" he is dropping the pebbles, the units for pebble-dropping should be pebbles/sec, right?

Let's draw a picture to solve our problem. The line graph in Figure 18-1 represents the distance traveled by the girl and boy and their wagon. I've marked it off in feet and included time labels in one-second intervals, which occur every three feet (they move 3 feet every second). The pebbles are provided in the required numbers (4 per foot). What do we see? If we count up the pebbles in an interval corresponding to 1 second, we find that there are 12. In other words, the boy has to dump the pebbles at a rate of 12 per second. We therefore have the answer to our question. To leave pebbles on the ground with a density of 4 pebbles/ft

Figure 18-1.

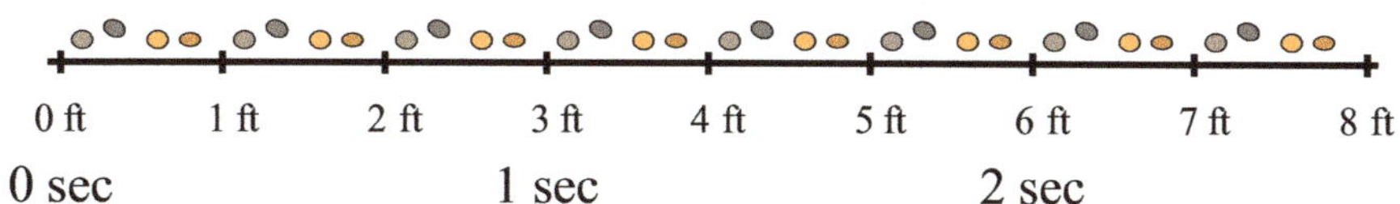

while his sister pulls him along at 3 ft/sec, the boy must drop the pebbles at the rate of 12 pebbles/sec.

Care to venture a guess how we would have arrived at this number without having to draw a picture and count the pebbles in each time interval? With a mathematical calculation, perhaps?

If we pay careful attention to the units of measurement, we can see that the following computation fits the bill:

$$4\,\frac{\text{pebbles}}{\text{ft}}\cdot 3\,\frac{\text{ft}}{\text{sec}} = 12\,\frac{\text{pebbles}}{\text{sec}}.$$

You can see how the units of length (feet) would cancel out and the end result would be expressed in pebbles per second, which is exactly what we wanted. Now let's try to generalize this by using some of Leibniz's notation for derivatives, which is the way calculus represents rates of change.

Write it like Leibniz

I suggest we use p for pebbles, x for distance traveled, and t for time. The girl's rate of travel is therefore the derivative of distance with respect to time, which can be written as

$$\frac{dx}{dt} = 3\text{ ft/ sec}.$$

The pebbles are scattered with a density of 4 pebbles/ft, so we can use Leibniz's notation to express that as a rate of change also:

$$\frac{dp}{dx} = 4\text{ pebbles/ft}.$$

Finally, we wanted to find out how fast the pebbles had to be dropped. Since that relates to time, the appropriate notation is

$$\frac{dp}{dt} = 12 \text{ pebbles/ sec}.$$

Although I showed you this notation earlier, we haven't really needed it till now because prime notation provided all the information we required. Now, however, we have a more complicated situation and prime just won't do. For example, what would p' mean? Sure, it's supposed to signify the rate of change of the variable p, but with respect to what? In our current problem, p has two different rates of change, one with respect to distance (the density of the pebbles per foot) and another with respect to time (the rate at which they're dropped). Leibniz notation may be a bit more cumbersome than prime notation, but it's quite explicit about what's going on and avoids all ambiguity.

Furthermore, check out what happens when we put the three Leibniz-notation derivatives together, rewriting the formula to reflect that 12 pebbles/sec = (4 pebbles/ft)(3 ft/sec):

$$\frac{dp}{dt} = \frac{dp}{dx} \cdot \frac{dx}{dt}.$$

This equation is one of the forms in which we state the chain rule. (Notice how much it looks like a calculation with regular fractions, as if the dx's cancel. They don't, exactly, but that's why Leibniz wrote it that way.) The chain rule is what we need to use whenever we get beyond the simple $y = f(x)$ situation. The girl-boy-pebble problem had three variables, not just two. We knew how p changed with respect to x and, in turn, we knew how x changed with respect to t, but what we really wanted to know was how p changed with respect to t. Mathematicians view each variable as a link in a chain, hence the name given to the rule. We could have situations with many more links, but the version we have now will be enough for our purposes. What we need now are some more general examples, by which I mean using formulas instead of specific numbers.

Composition 1A

Suppose we know that $p = x^2$ and $x = 3t + 1$. I picked these because they are neither too simple nor too complicated. (This is a new example, not related to the children and their pebbles.) We can certainly use the chain rule on them to find out the rate of change of p with respect to t. Here we go, working out the first two derivatives:

$$p = x^2, \text{ so } \frac{dp}{dx} = 2x,$$
$$x = 3t + 1, \text{ so } \frac{dx}{dt} = 3.$$

If we plug our two results into the chain rule formula, we get our new derivative for p with respect to t:

$$\begin{aligned}\frac{dp}{dt} &= \frac{dp}{dx} \cdot \frac{dx}{dt} \\ &= 2x \cdot 3 \\ &= 6x.\end{aligned}$$

Thanks to the chain rule, we now have an expression for the rate of change of p with respect to t.

Composition 1B

Earlier we discussed how mathematicians are lazy. Perhaps you're an aspiring mathematician and it's already occurred to you that all this new stuff isn't really necessary. Couldn't we do it with less work? Since we said that $p = x^2$ and $x = 3t + 1$, why not just put it all together? We could simply plug the value of x into the expression for p:

$$p = x^2 = (3t + 1)^2 = 9t^2 + 6t + 1.$$

Now that p is expressed purely in terms of t, I could actually use prime notation as shown on the next page.

$$\begin{aligned}\frac{dp}{dt} &= p' \\ &= (9t^2 + 6t + 1)' \\ &= 9 \cdot 2t + 6 + 0 \\ &= 18t + 6.\end{aligned}$$

Our example had a linear function, $x = 3t + 1$, being plugged into the second-degree function $p = x^2$. You may recall from algebra that plugging one function into another is called *composition*. The chain rule is a way of finding derivatives of composite functions.

You could be forgiven for thinking that was a lot easier than using the chain rule (especially if you ignore the work it took to do the substitution of $x = 3t + 1$ into the expression for p and squaring), but most of the time the chain rule is actually a short cut. We're still dawdling and writing down a lot of steps for the chain rule because it's new to us. It'll get faster.

By the way, why are our results different? The chain rule says that $\frac{dp}{dt} = 6x$, while the version we got by plugging in and priming was $18t + 6$. Look at both of them and see if you can recognize why they're actually the same. It might help if you remind yourself that $x = 3t + 1$ and multiply it by 6: $6x = 6(3t + 1) = 18t + 6$. Yeah, they're the same, all right.

Here's another way of looking at the chain rule computation we just did, but this time in terms of compositions with function notation. We know that $p = f(x) = x^2$, which $x = g(t) = 3t + 1$. When we put them together, we get

$$p = f(x) = f(g(t))$$

that is, one function is nested inside the other.

Note also that

$$\frac{dp}{dx} = f'(x) = 2x \quad \text{and} \quad \frac{dx}{dt} = g'(t) = 3$$

so when we rewrite the chain rule using these functions, we get

$$\frac{dp}{dt} = \frac{dp}{dx} \cdot \frac{dx}{dt} = f'(x) \cdot g'(t).$$

It's just a product of the derivatives of the two functions, right? That's what Leibniz originally expected for the product rule (see Chapter 21), but in reality the product of the derivatives of the two functions gives the derivative of their composition, not their product. People who want to write the chain rule with function notation and primes (perhaps because they don't like Leibniz notation as much as I do) often write it this way:

$$f(g(t))' = f'(g(t)) \cdot g'(t).$$

It's not as pretty in this format, but it does help us see one very important point: To find the derivative of a composite function, you just (a) take the derivative of the outer function and (b) multiply by the derivative of the inner function. Since I've beaten you over the head with our current example often enough, let me show you a trigonometric case that provides a wonderful picture of what's going on.

A composition with trig

We know that the derivative of the sine function is the cosine function: $(\sin x)' = \cos x$. What happens if we change the sine function by sticking something inside it? For example, what is the derivative of $y = \sin(2x)$? If we look at Figure 18-2, the graphs of $y = \sin x$ and $y = \sin(2x)$ tell us all we need to know. Keep your eye on the point where I'm drawing tangent lines: the place to the right of the origin where the curve first cuts across the x axis.

Figure 18-2.

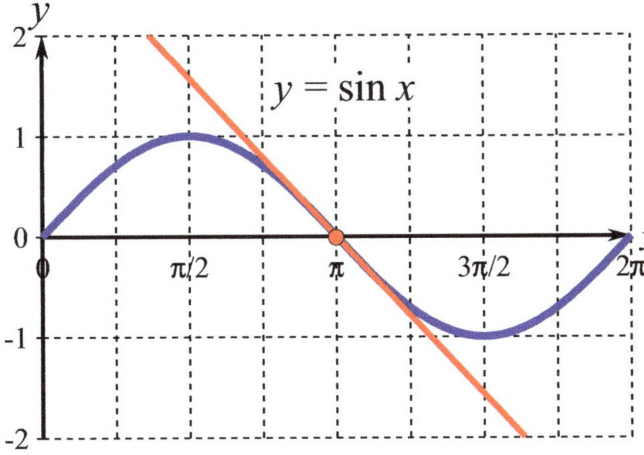

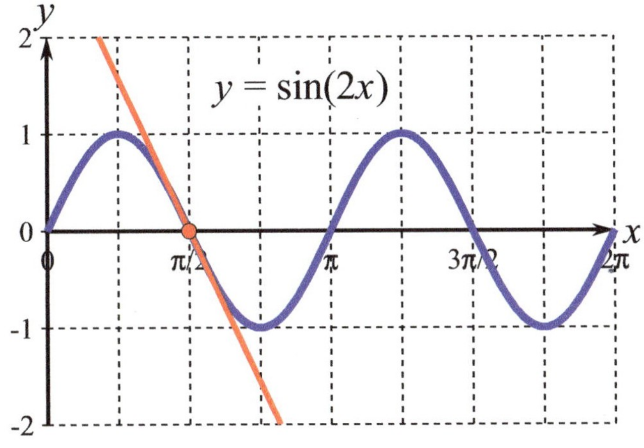

If you look at the point $(\pi, 0)$ on the curve $y = \sin x$, you know the slope of the tangent at that point has to be given by cos(π) = –1. I've drawn in the tangent line and you can see that $m = -1$ looks good. But for the curve $y = \sin(2x)$, the curve is compressed horizontally. The sine curve wiggles twice as fast with a 2 in it. What is the slope of the tangent line at the point (π/2, 0)? According to the chain rule, the derivative is

$$\begin{aligned} y' &= (\sin(2x))' \\ &= \cos(2x)\cdot(2x)' \\ &= \cos(2x)\cdot 2 \\ &= 2\cos(2x). \end{aligned}$$

As you can see, we took the derivative of sine, which gave us cosine, and then we multiplied by the derivative of what's inside, which gave us a factor of 2. The slopes on the second curve have all doubled with respect to the first curve, but that makes perfectly good sense because slope is a measure of rate of change—and the rate of change has indeed doubled, as the graphs show us visually. If we follow through by computing the slope at (π/2, 0) for the curve $y = \sin(2x)$, we get

$$\begin{aligned} y'\big|_{\pi/2} &= 2\cos 2x\big|_{\pi/2} \\ &= 2\cos\left(2\cdot\frac{\pi}{2}\right) \\ &= 2\cos(\pi) \\ &= 2(-1) = -2. \end{aligned}$$

Just as we were led to expect, the slope is $m = -2$.

The last trig example I'd like to show you in this chapter is even more ambitious than merely doubling what's inside the sine. This time I'm replacing x with x^2. In this case, the rate at which the sine curve oscillates up and down does not just double, it grows faster and faster and faster. Just like x^2.

What is the derivative of $y = \sin(x^2)$? Let's apply the chain rule:

$$\begin{aligned} y' &= (\sin(x^2))' \\ &= \cos(x^2)\cdot(x^2)' \\ &= \cos(x^2)\cdot(2x) \\ &= 2x\cos(x^2). \end{aligned}$$

Recall that I want to look at the point in Figure 18-3 where the curve first crosses the x axis. That means we need $\sin(x^2) = 0$, so that the whole function is 0. Well, $\sin\pi = 0$, so we must have $x^2 = \pi$ and therefore $x = \sqrt{\pi}$. Using the derivative formula we just found, we can find the slope of the tangent line to the curve $y = \sin(x^2)$ at the point $(\sqrt{\pi}, 0)$:

$$\begin{aligned} y'|_{\sqrt{\pi}} &= 2x\cos(x^2)|_{\sqrt{\pi}} \\ &= 2\sqrt{\pi}\cos(\pi) \\ &= 2\sqrt{\pi}(-1) \\ &= -2\sqrt{\pi} \approx -3.5449. \end{aligned}$$

The slope should be about –3.5. Check the graph in Figure 18-3 to make sure the result makes sense. It matches quite nicely.

Thanks to the chain rule, we can now find the derivatives of functions of functions, one link at a time.

Figure 18-3.

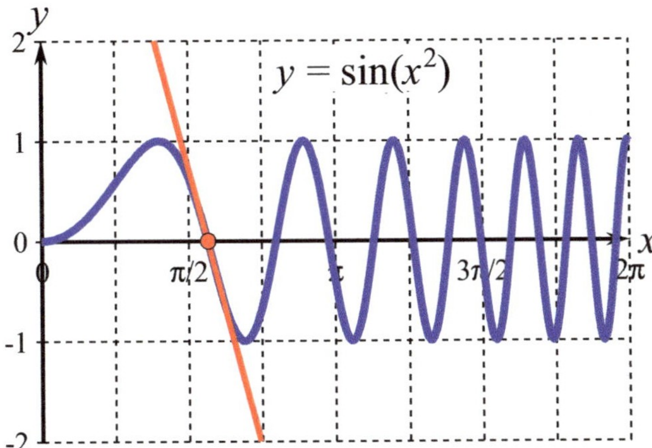

19 Transcendental Mediations on Functions

A little log rolling, and unrolling

We have been doing a lot of work with polynomials. They were the functions we began with, and they keep returning in examples. The great thing about polynomials is that they're nothing more than arithmetic with some variables mixed in: Take some powers of x, multiply them by whatever numbers you like, and add, subtract, or multiply them together. (To make things even nicer, we avoid division.) Polynomials are great. When we toss some roots (square roots, cube roots, etc.) into the mix, we end up with the class of functions called *algebraic functions*. Yes, indeed, they are the functions we know from our algebra classes.

Trigonometric functions are different. These functions are called *transcendental* because their results cannot be expressed in terms of the algebraic operations we know from polynomials and roots. The trig functions were our first examples of transcendental functions, but the trig functions are by no means alone in this category. There are some very important non-trig transcendental functions. I want to introduce you to two of them. They'll complete the set of functions we'll be using in the remainder of our stroll through calculus. I'll begin with *exponential functions*.

We use the term *power function* to refer to something of the form x^n, where the variable x is in the base of the expression and the constant n is in the exponent. (A power function is definitely algebraic, being the basis of the polynomials.) An *exponential function* is the opposite, with the variable in the exponent and a constant in the base. (By contrast with the power function, it is transcendental.) Out of many possible examples, I'd like to begin with $f(x) = 2^x$. This is sometimes called the base-2 exponential function, but more often people just say "two to the x" when they want to refer to it.

You may have heard "exponential" used in a nonmathematical context and picked up a pretty good idea of how exponential functions behave from context. For example, a news article might report that the national debt is "growing exponentially." The implication is clear: Exponential growth is pretty big. Here's a table of values for $f(x) = 2^x$ and an accompanying graph to show how rapidly this particular exponential function increases.

x	–2	–1	0	1	2	3	4	5	6
2^x	1/4	1/2	1	2	4	8	16	32	64

As demonstrated by Figure 19-1, the graph of $f(x) = 2^x$ goes rocketing upward. Let's find out exactly how fast the curve rises by finding the derivative. Since we've never seen a function like this before, we need to go all the way back to the basic definition of the derivative:

$$\begin{aligned} f'(x) &= \lim_{\Delta x \to 0} \frac{f(x+\Delta x) - f(x)}{\Delta x} \\ &= \lim_{\Delta x \to 0} \frac{2^{x+\Delta x} - 2^x}{\Delta x} \\ &= \lim_{\Delta x \to 0} \frac{2^x 2^{\Delta x} - 2^x}{\Delta x} \quad \text{(recall that } 2^{m+n} = 2^m 2^n\text{)} \\ &= \lim_{\Delta x \to 0} \frac{2^x(2^{\Delta x} - 1)}{\Delta x} \\ &= 2^x \lim_{\Delta x \to 0} \frac{2^{\Delta x} - 1}{\Delta x}. \end{aligned}$$

Figure 19-1.

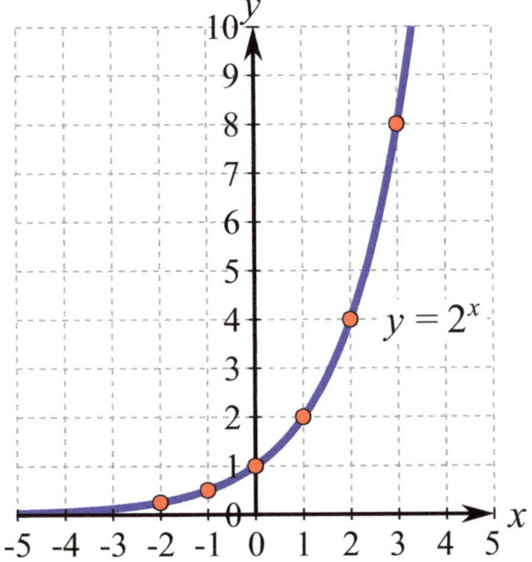

We factored out the 2^x because the limit has nothing to do with it (the limit is all about Δx), but now we're stuck. The limit that's left is *not* a plug-in. Go ahead and try. If we replace Δx with 0 as shown on the next page, we get a most unfortunate result.

$$2^x \lim_{\Delta x \to 0} \frac{2^{\Delta x} - 1}{\Delta x} = 2^x \cdot \frac{2^0 - 1}{0}$$
$$= 2^x \cdot \frac{1 - 1}{0}$$
$$= 2^x \cdot \frac{0}{0},$$

which I'm afraid is just meaningless.

Let's try a numerical approach instead. If the limit exists, we should be able to get an idea of its value by plugging in some very small numbers for Δx. Here's a table of the results:

Δx	0.1	0.01	0.001	0.0001	0.00001
$\frac{(2^{\Delta x} - 1)}{\Delta x}$	0.7177	0.6956	0.6934	0.6932	0.6932

These numbers strongly suggest that the limit is close to 0.693, in which case we would have the formula

$$(2^x)' = 2^x(0.693).$$

That is, 2^x is pretty much its own derivative, except that an extra factor of 0.693 is required. This is a peculiar situation, since there is no apparent reason that such a weird factor as 0.693 should show up. However, it does appear to work. If I plug $x = 0$ into $f'(x) = 2^x(0.693)$, I get $f'(0) = 2^0(0.693) = 0.693$, which should therefore be the slope of the graph of $f(x) = 2^x$ when $x = 0$. If I draw a tangent line with slope 0.693 at the point (0, 1), it looks like a good fit. See Figure 19-2.

Figure 19-2.

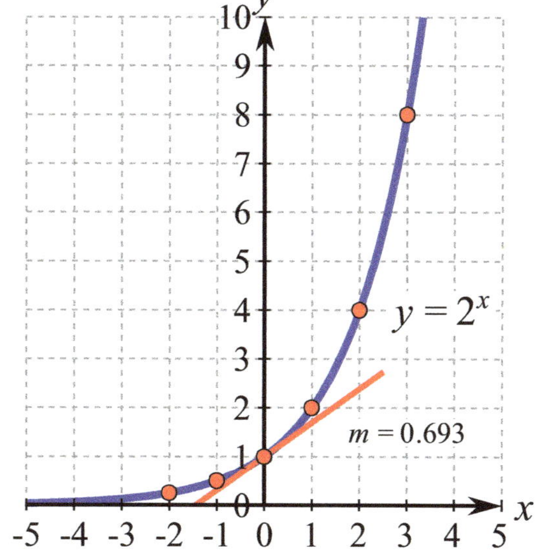

Far off base

Our derivative for 2^x is not pretty, but it seems to work. What would happen if we try other bases? If, for example, we worked through the limit of the difference quotient for 3^x, everything would be exactly the same as for 2^x, except that we would be left looking at

$$(3^x)' = 3^x \lim_{\Delta x \to 0} \frac{3^{\Delta x} - 1}{\Delta x}.$$

Right? (You might want to check this claim by filling in the details yourself.) Then we'd see that 3^x is pretty much its own derivative, except for its own peculiar extra factor. I've done the calculator work for several additional base choices and listed the results in a table for you. Each different choice of base produces a different extra factor. Here they are:

a	$\lim_{\Delta x \to 0} \frac{a^{\Delta x} - 1}{\Delta x}$	$(a^x)'$
2	0.693	$2^x(0.693)$
3	1.099	$3^x(1.099)$
4	1.386	$4^x(1.386)$
5	1.609	$5^x(1.609)$
10	2.306	$10^x(2.306)$

Too bad about base 3. The multiplying factor in its derivative was 1.099. If only it had been exactly 1.000! That would have been the ultimate in convenience. It would also mean that the exponential function would be its own derivative, which would have been quite remarkable. But no such luck. The situation is even worse for base 4, because we see the multiplier factor gets bigger as larger bases are chosen.

Are you thinking what I'm thinking? Base 3 was just a little bit too big to get a multiplier factor of 1. Base 2 was apparently too small. There ought to be a number, just a little smaller than 3, for which the multiplier would be exactly 1. Such a number would be our perfect exponential base. Check out the results on the next page.

a	$\lim_{\Delta x \to 0} \frac{a^{\Delta x} - 1}{\Delta x}$	$(a^x)'$
2.7	0.993	$2.7^x(0.993)$
2.8	1.030	$2.8^x(1.030)$
2.9	1.065	$2.9^x(1.065)$

We are homing in on the perfect base. It must be between 2.7 and 2.8—a number really close to 2.7. What kind of number is this elusive perfect exponential base?

Mathematicians have now tracked this number down to a huge number of decimal places. Leonhard Euler, a Swiss genius who is one of the most famous mathematicians to follow in the footsteps of Newton and Leibniz, decided to call the number e and worked out its value to 23 decimal places:

$$e \approx 2.71828182845904523536028,$$

which is probably a little more than you or I will need.

We see that e is a bizarre number, but mathematicians are extremely fond of it. Remember why it was tracked down. If you consider the exponential function $f(x) = e^x$, then $f'(x) = e^x$. It is its own derivative! Only exponentials based on e behave this way. In consequence of this fact, mathematicians hardly ever work with any other base for their exponential functions: e^x rules. We even refer to e^x as the *natural* exponential function, as if all other bases are not.

The messy work we did getting close to e (we didn't find it ourselves, you know; I punted and let Euler take over) shows us that the use of e is a trade-off. The exponential function has a great derivative (itself!) when based on e. In exchange for this amazing derivative, we accept a transcendental number as the base. This is just another example of the law of conservation of mathematical difficulty: If you make things nicer in one place, some complication is sure to pop up somewhere else. We do the best we can and make our choices.

By the way, you may see that e^x has a prominent place on the keyboard of your scientific calculator. Most math students get introduced to the natural exponential function in Algebra 2, when it's used to compute population growth or the decay of radioactive isotopes. (Remember?) The natural companion of the natural exponential function is the *natural logarithm function*, which is denoted "ln x." The appropriate calculator key will carry a label like that, or possibly just "LN." If you find this key and use it to compute ln 2 (the natural logarithm of 2), you'll get a small surprise. Perhaps. Now compute ln 3. See what's going on?

We're getting the numbers from the table of multiplicative factors that we worked out earlier. With the assistance of the LN key, we have (to four decimal place accuracy) ln 2 = 0.6931, ln 3 = 1.0986, ln 4 = 1.3863, etc. Recognizing this fact permits us to add a new rule to our list of derivatives:

$f(x)$	$f'(x)$
e^x	e^x
a^x	$a^x(\ln a)$

That is, for example, $(2^x)' = 2^x(\ln 2)$. That looks a little better than the version with 0.693, but it's still not as nice as our *natural* result, $(e^x)' = e^x$.

We've settled the issue of the derivative of the exponential function. Now let's find out how to take the derivative of a logarithmic function, beginning with ln x itself.

Postscript

Isn't it odd that two of the most important numbers in math are so close to 3? We know that π is just a little larger and e is just a little smaller.

Perhaps there is some alternate universe in which π and e have the same value and that value is 3. Surely math students would have it easier in that universe. Unfortunately, the law of conservation of mathematical difficulty probably means that something else would become more complicated. (In case you didn't suspect, permit me to confess that there is no real law of conservation of mathematical difficulty. I made it up.)

20 As Easy as Falling Off a Logarithm

It's not a caber-tossing contest

I mentioned in the previous chapter that e^x, the natural exponential function, is normally introduced in Algebra 2. Its usual companion is $\ln x$, the natural logarithm function. These two functions go together because they are inverses of each other. It turns out that this relationship is the key that will enable us to find the derivative of the natural logarithm.

Let's remind ourselves what inverse functions do: They cancel each other out. The functions $f(x)$ and $g(x)$ are inverses of each other if $f(g(x)) = x$ and $g(f(x)) = x$. That is, their compositions have no effect on x. Whatever value of x you plug in, you just get it back. One of the best examples of a pair of inverse functions is provided by the dynamic duo of the squaring function and the square root function. What is the square of 3? It's 9. What is the square root of 9? It's 3. You end up where you started. We began with 3 and ended with 3.

Try it the other way around: What's the square root of 9? It's 3. What's the square of 3? It's 9. This time we started with 9 and ended up with 9. In symbols, we can write our two examples this way:

$$\sqrt{3^2} = 3 \text{ and } (\sqrt{9})^2 = 9.$$

Figure 20-1 shows us how the graphs of $y = x^2$ and $y = \sqrt{x}$ relate to each other. As I explained, the squaring function turns 3 into 9, and the square root function turns 9 back into 3. That's why one graph has (3, 9) on it and the other graph has (9, 3). The graphs of inverse functions are always mirror images of each other (the "mirror" is the diagonal line $y = x$, as I think you can see from Figure 20-1).

Figure 20-1.

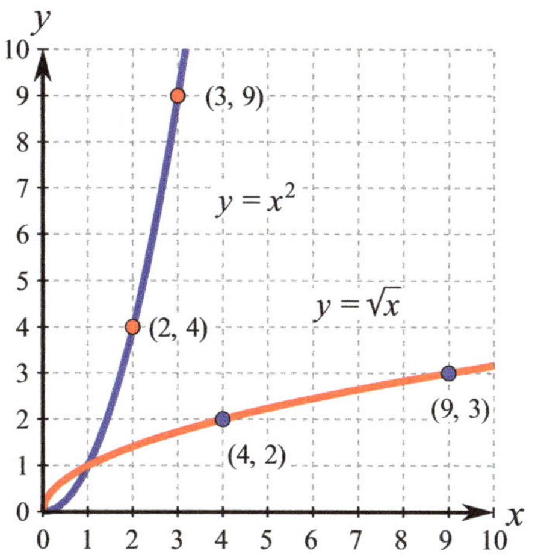

One at a time, please

Say, doesn't 9 have two square roots? After all, both 3^2 and $(-3)^2$ equal 9. That's true, but the square-root key on your calculator—the one marked with a $\sqrt{\ }$ —gives you only one answer (namely, 3) when you ask it for the square root of 9. The keys on a calculator are called *function* keys for a good reason: functions produce at most one output value for each input value, never two or three or more. Otherwise we run into ambiguities and confusion in our calculations. In the case of square roots, for example, the function $\sqrt{x}$, which gives only nonnegative results, is the inverse function for the function x^2, provided that we restrict the input (the *domain*) for x^2 to nonnegative numbers.

Let's express this property of inverse functions more generically: Suppose your inverse functions are called $f(x)$ and $g(x)$. Any time you have $f(a) = b$, then $g(b) = a$. That's because the inverse functions $f(x)$ and $g(x)$ undo each other. See Figure 20-2 for an example. You'll also notice from the figure that the slope at (a, b) must be $f'(a)$, and the slope at (b, a) must be $g'(b)$. Furthermore, since the graphs of $f(x)$ and $g(x)$ involve swapping the x and y coordinates of each point, it follows that—and this is important!—*rise and run get swapped* for every tangent line. In brief, the slopes at corresponding points of the graphs must be reciprocals of each other:

$$g'(b) = \frac{1}{f'(a)}.$$

If, for example, in Figure 20-2 the slope at (a, b) is $\frac{3}{2}$ (and I did draw it that way), the slope at (b, a) will have to be $\frac{2}{3}$. So now we know that inverse functions have reciprocal slopes.

Figure 20-2.

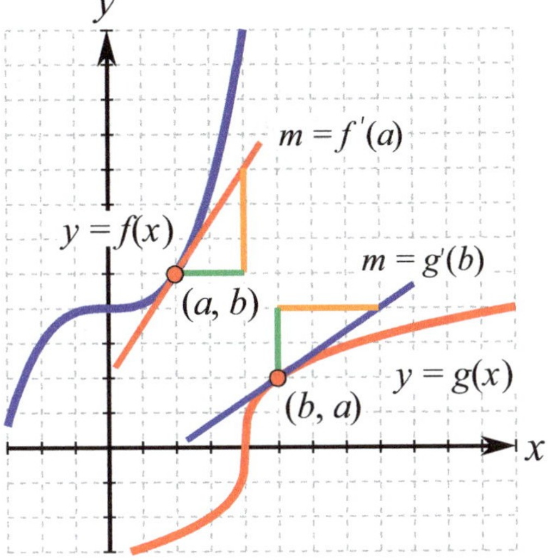

Let's check this out in the case of the exponential function and the logarithmic function (base e, of course). In Figure 20-3, we see their graphs. The graphs are mirror images of each other, with (0, 1) on the exponential curve and (1, 0) on the logarithmic curve (for example). We also see that $(1, e)$ is on the exponential curve and $(e, 1)$ is on the logarithmic curve. In general, if $e^a = b$, then the point (a, b) is on the exponential curve. Then (b, a) must be on the logarithmic curve, which implies that $\ln b = a$. We already know that $\ln 2 = 0.6931$, so the points (0.6931, 2) and (2, 0.6931) provide a nice numerical example we can check with our calculators. (Try out some other points, too, just to make it clear.)

Figure 20-3.

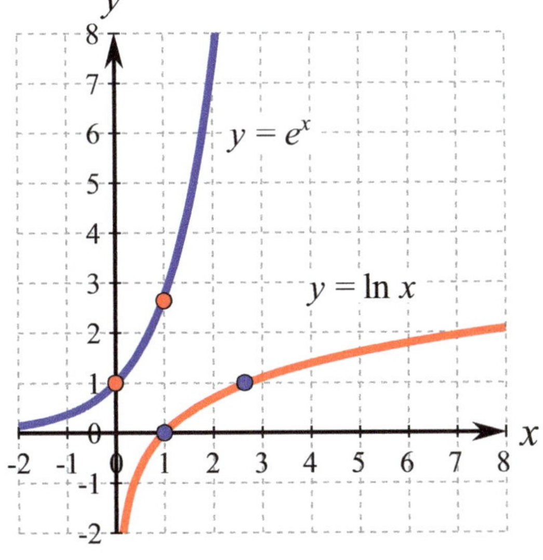

Let's apply what we know about the derivatives of inverse functions to the case of the natural exponential and natural logarithmic functions. Since $y = e^x$ leads to $y' = e^x$, the slope at any point (x, y) on the exponential curve is $m = y' = y$. I've drawn a bunch of these tangents in Figure 20-4. The slope at (0, 1) on the exponential curve is $m = 1$, the slope at $(\ln 2, 2)$ is $m = 2$, the slope at $(\ln 3, 3)$ is $m = 3$, and the slope at $(\ln 4, 4)$ is $m = 4$. What does this tell us about the slopes at the corresponding points on the logarithmic curve?

Well, recall that the slopes of the logarithmic curve will be reciprocals. For instance, the reciprocal of 2 is $\frac{1}{2}$, so at the point $(2, \ln 2)$, the slope of the tangent line will be $\frac{1}{2}$. Likewise, the reciprocal of 3 is $\frac{1}{3}$, so at the point $(3, \ln 3)$, the slope will be $\frac{1}{3}$. In general, the slope at $(x, \ln x)$ must be $\frac{1}{x}$. Let's write it as a formula:

$$(\ln x)' = \frac{1}{x}.$$

The derivative of the natural logarithm function is the reciprocal function. This result is not as simple as the one for the natural exponential, but it's still rather neat and tidy. Once again, e proves its worth as a base for our natural functions.

Figure 20-4.

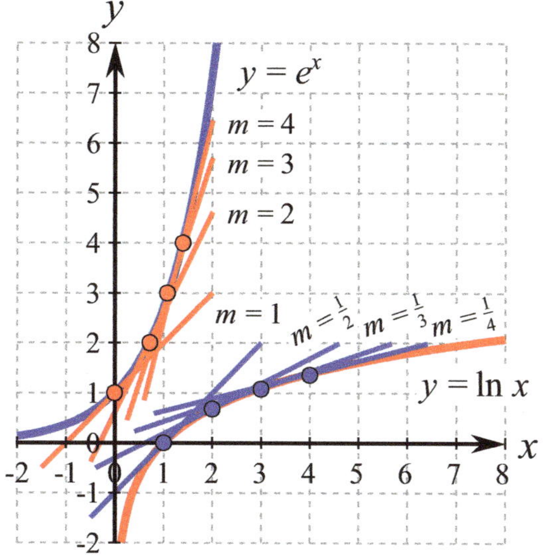

The power rule made complete

There are some immediate implications from the formulas we worked out in this and the preceding chapter. Every derivative formula is automatically an integral formula, too, so we may now write two new antiderivatives:

$$\int e^x = e^x + C,$$
$$\int \frac{1}{x} = \ln x + C.$$

Both of these are very useful in mathematical applications, and the first one is the simplest antiderivative you could ever hope for, but the second antiderivative is something we should be especially pleased about. We all remember the power rule for integrals:

$$\int x^n = \frac{1}{n+1} x^{n+1} + C.$$

Although we didn't make a big deal about it at the time, it's clear that this rule cannot work if $n = -1$. That would lead to division by zero in the result. Let's confirm this:

$$\begin{aligned}\int \frac{1}{x} &= \int x^{-1}\\ &= \frac{1}{-1+1}x^{-1+1} + C\\ &= \frac{1}{0}x^{0} + C.\end{aligned}$$

This is a bit of a pickle. The power rule failed quite dramatically for $n = -1$. Up to this point, therefore, we had no antiderivative for the reciprocal function $\frac{1}{x}$. Thanks to the work we did in this chapter, however, that problem is nicely solved. Since

$$(\ln x)' = \frac{1}{x},$$

we can also say that

$$\int \frac{1}{x} = \ln x + C.$$

Here's another interesting fact. If we use the FTC to work out the definite integral of $\frac{1}{x}$ from 1 to e, we get

$$\begin{aligned}\int_1^e \frac{1}{x} &= \ln x\big|_1^e\\ &= \ln e - \ln 1\\ &= 1 - 0 = 1.\end{aligned}$$

That's right: e is the magic number for which the area (starting from $x = 1$, as shown in Figure 20-5) under the reciprocal curve $y = \frac{1}{x}$ is exactly 1. Sometimes e is defined in this way, but we went another route, discovering instead that e is the number that provides the best of all possible bases for exponentials and logarithms. It's a natural.

Figure 20-5.

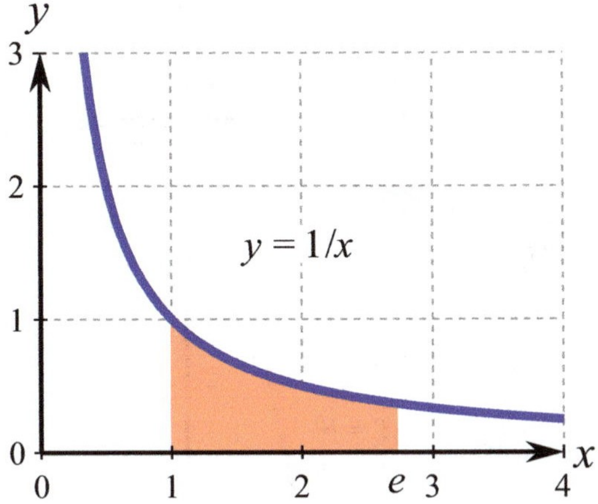

21 An Integral of Many Parts

Unraveling the product rule

The previous chapter completed our collection of basic functions: powers, polynomials, radicals, trigonometric, exponential, and logarithmic. We call these *elementary functions*, and they suffice for the vast majority of calculus applications. We know the derivatives of all of them, including their sums, differences, products, quotients, and compositions. Their antiderivatives, however, are not always as easy to find.

You're already aware that every derivative formula is also an antiderivative formula. For example, the fact that $(\tan x)' = (\sec x)^2$ means that $\int (\sec x)^2 = \tan x + C$. But many gaps remain in our collection of antiderivatives. Do we know one for $\tan x$? No. Do we know one for $\ln x$? Not yet.

What you may not have realized, however, is that derivative rules like the product rule and chain rule can also be turned into antiderivative rules. In this chapter, I'm going to show you how to turn the product rule into what mathematicians call integration by parts. In the next chapter, we'll similarly unravel the chain rule.

To begin, let's recall the product rule:

$$(f(x)g(x))' = f(x)g'(x) + f'(x)g(x).$$

What happens if we apply the integral sign to both sides of this equation? We get the results on the next page.

$$\int (f(x)g(x))' = \int [f(x)g'(x) + f'(x)g(x)]$$
$$= \int f(x)g'(x) + \int f'(x)g(x).$$

Of course, we know that

$$\int (f(x)g(x))' = f(x)g(x) + C$$

because the integral cancels out the derivative. We now have

$$f(x)g(x) + C = \int f(x)g'(x) + f'(x)g(x).$$

With a little algebra, we can rearrange this by isolating the first integral on the right side. We end up with

$$\int f(x)g'(x) = f(x)g(x) - \int f'(x)g(x) + C.$$

This is the formula for "integration by parts," as I mentioned earlier. It's traditionally written without the "+ C" on the right side because the presence of the integral sign reminds us to include it if we actually use the formula. Let me show you how it works.

What is the antiderivative of the product $x \cos x$? To use integration by parts, we need to break the integrand of

$$\int x \cos x$$

in two. One piece is supposed to be $f(x)$, and the other part is supposed to be $g'(x)$. How shall we choose? Let's try the obvious thing and let $f(x) = x$ and $g'(x) = \cos x$. We then know that $f'(x) = 1$ and $g(x) = \int g'(x) = \int \cos x = \sin x + C$. We don't really need the C, so I'll leave it off as we plug into the formula for integration by parts, substituting our values for $f(x)$, $g(x)$, and their derivatives on the next page.

$$\int f(x)g'(x) = f(x)g(x) - \int f'(x)g(x)$$
$$\begin{aligned}\int x \cos x &= x \sin x - \int 1 \cdot \sin x \\ &= x \sin x - \int \sin x \\ &= x \sin x - (-\cos x) + C \\ &= x \sin x + \cos x + C.\end{aligned}$$

According to the formula for integration by parts, the antiderivative for $x \cos x$ is $x \sin x + \cos x + C$. How can we check this? Easy. Let's take the derivative of this result to see if we end up with $x \cos x$ again:

$$\begin{aligned}&(x \sin x + \cos x + C)' = (x \sin x)' + (\cos x)' + C' \\ &= x(\sin x)' + (x)'(\sin x) + (-\sin x) + 0 \\ &= x \cos x + 1 \cdot \sin x - \sin x \\ &= x \cos x.\end{aligned}$$

It works! I hope you weren't too surprised that we had to use the product rule. After all, integration by parts involves using the product rule backward, so it stands to reason that its results will have products in them.

Working with nothing

My second example of integration by parts is like a magic trick. We're going to conjure one of our functions out of thin air. Our goal is to find an antiderivative for $\ln x$. But how can we possibly fit

$$\int \ln x$$

into the pattern for integration by parts? That technique specifically requires that we break the integrand into two separate factors, one called $f(x)$ and the other called $g'(x)$. Our current problem has nothing in the integrand except for $\ln x$, all by itself.

A neat sleight of hand does the job. If we let $f(x) = \ln x$ and $g'(x) = 1$, then $f(x)g'(x)$ does indeed equal $\ln x$. I don't know who was the first person to think of using 1 as one of the factors, but it's a very clever move. We then have $f'(x) = (\ln x)' = \frac{1}{x}$ and $\int g(x) = \int g'(x) = \int 1 = x + C$. (I'm going to ignore that "+ C" again; it'll pop up on its own again later.) Plugging into the formula for integration by parts gives us

$$\begin{aligned}
\int f(x)g'(x) &= f(x)g(x) - \int f'(x)g(x) \\
\int \ln x &= (\ln x)(x) - \int \frac{1}{x} \cdot x \\
&= x \ln x - \int 1 \\
&= x \ln x - x + C.
\end{aligned}$$

There it is—the antiderivative of $\ln x$. You can check this by taking the derivative if you like.

22 The Anti-Chain Rule

Rise up, integrals of the world

The chain rule gives us the recipe for finding the derivative of a composite function—that is, a function of a function. As you'll recall, we end up with the derivatives of the two functions multiplied together, like this:

$$(f(g(x)))' = f'(g(x))g'(x).$$

If we apply an integral sign to both sides, we're now looking at

$$\int (f(g(x)))' = \int f'(g(x))g'(x).$$

The left side is easy to simplify, since the integral sign just cancels out the derivative, so we're just going to get $f(g(x)) + C$. If we plug this in and rearrange the equation, we end up with a useful formula:

$$\int f'(g(x))g'(x) = f(g(x)) + C.$$

An example will help us see how this works. Suppose we want to compute

$$\int (\sin x)^2 \cos x.$$

Does this fit the pattern of the anti-chain rule? The first thing to notice is that we should probably choose $g(x) = \sin x$, because that's the function inside the square in the integrand. Furthermore, $g'(x) = (\sin x)' = \cos x$, so the necessary pieces are in place. So what's $f(x)$? Looking at the integrand, we see that $f'(x)$ has to be the squaring function; that is, $f'(x) = x^2$ means that $f'(g(x)) = (g(x))^2 = (\sin x)^2$, just as required. Furthermore,

$f'(x) = x^2$ implies that $f(x)$ has to be the antiderivative of x^2, namely,

$$f(x) = \int f'(x) = \int x^2 = \frac{1}{3}x^3 + C.$$

When we plug the expressions for $f(x)$ and $g(x)$ and their derivatives into the anti-chain rule formula, we get a nice result:

$$\begin{aligned}\int f'(g(x))g'(x) &= f(g(x)) + C\\ \int (g(x))^2 g'(x) &= \frac{1}{3}(g(x))^3 + C\\ &= \frac{1}{3}(\sin x)^3 + C.\end{aligned}$$

We use the chain rule, of course, to check our answer. Do we get the original integrand when we take the derivative of this result? We do:

$$\begin{aligned}&\left(\frac{1}{3}(\sin x)^3 + C\right)' = \frac{1}{3}\cdot 3(\sin x)^2(\sin x)' + 0\\ &= (\sin x)^2 \cos x.\end{aligned}$$

Okay, it worked out just fine, but you might think it was quite a stretch of the imagination to figure out $f(x)$ from the mess we had in the original integrand. Could there be a better way?

Let *u* do it

Leibniz gets to take one more big bow right here. Mathematicians refer to the process of using the anti-chain rule as *integrating by substitution*. Since you didn't really see any substitution in our example, you could be forgiven for thinking integration by substitution isn't a very good name. In fact, it's a perfect name, and I will soon try to justify that claim. To do a proper job, though, I'm going to have to step back a few paces and take a run at it.

I want to start with a teaser: Do you know what $\int 1$ is? Would you say

$$\int 1 = x + C,$$

or would you say something else, like maybe $t + C$? How would you know which one to choose? In general, we've depended on context. If we've been using x as our variable, we'd say the antiderivative of 1 is $x + C$. If we've been talking in terms of time and using t, then we'd say the antiderivative of 1 is $t + C$. It hasn't really been a problem so far.

Leibniz, however, was not one to stand for such ambiguity. He tacked a symbol on to his integrals to make it perfectly clear what variable to use. It looked like this in the x case:

$$\int 1\,dx = x + C.$$

So where did this idea come from? The dx looks like a refugee from Leibniz's $\frac{dy}{dx}$ notation for derivatives, *and that's exactly what it is* (as we'll see in a moment). Of course, most of the time it's perfectly clear what the variable is, so the extra dx (or whatever) seems redundant, but it's going to turn out to be quite powerful. Here are some examples of Leibniz's full integral notation at work, looking pretty much like things we've seen before—with just a little extra:

$$\int x^2\,dx = \frac{1}{3}x^3 + C$$
$$\int (\cos t)dt = \sin t + C$$
$$\int \frac{1}{u}du = \ln u + C$$
$$\int 5\,dz = 5z + C$$
$$\int \xi^3\,d\xi = \frac{1}{4}\xi^4 + C.$$

As you can see, only the case where the integrand was only 5 really depended on the extra bit to tip us off that the variable was z instead of something else. I also must apologize for that last example, the integral with the squiggle character (it's the lowercase Greek letter *xi*), but I really wanted to make sure that you understood that Leibniz's integral notation can be used with any variable we want. Any variable at all. I'll stick to ordinary letters for variables in what follows.

To see how Leibniz cooked this up (and why), remember that he used $\frac{dy}{dx}$ for the derivative of y with respect to x. If $y = f(x)$, we could write this out as

$$\frac{dy}{dx} = f'(x).$$

But Leibniz did not stop there. He really liked to think of dy and dx as tiny numbers ("infinitesimals") whose ratio was the slope of the tangent line to the graph of $f(x)$. He therefore felt free to break them apart and rewrite the previous derivative equation as

$$dy = f'(x)dx.$$

He called dy and dx "differentials," which is why the derivative portion of calculus is often referred to as *differential calculus*. (The other part, of course, is *integral calculus*.) Differentials are as easy to compute as derivatives. In fact, they're just slight variants of each other. Here's a table with a few examples to contrast them:

Function	***Derivative***	***Differential***
$y = x^2$	$\frac{dy}{dx} = 2x$	$dy = 2x\, dx$
$u = \sin x$	$\frac{du}{dx} = \cos x$	$du = \cos x\, dx$
$y = \ln t$	$\frac{dy}{dt} = \frac{1}{t}$	$dy = \frac{1}{t}dt$
$z = w^2 + w$	$\frac{dz}{dw} = 2w + 1$	$dz = (2w+1)\, dw$

Nothing to it.

Now let me make good on my promise to show you the power of Leibniz's differential notation when it comes to integration by substitution (the anti-chain rule). Leibniz liked to think of d as a special differential symbol that was useful to create derivatives and $\int$ as the special antiderivative symbol that cancels out the d. In other words, $\int dy = y + C$. So what happens if we apply the integral sign to both sides of the differential equation $dy = f'(x)$? We get

$$\int dy = \int f'(x)dx$$
$$y + C = \int f'(x)dx.$$

But wait a minute: We know that $y = f(x)$, so our result becomes

$$f(x) + C = \int f'(x)dx.$$

Okay, this is hardly a news flash, because we've known that integration cancels derivatives ever since we learned the Fundamental Theorem of Calculus, but at least now we see how Leibniz worked out his notation. And we're ready to apply the lessons of Leibniz notation to our earlier example,

$$\int (\sin x)^2 \cos x \, dx,$$

which I am now writing with the terminal differential dx. The traditional substitution variable is u, and the procedure calls for us to find the "inside" function in the integrand and call it u. As we discussed earlier, it's rather clear that $\sin x$ is the function that is inside another function (namely, the squaring function). So let $u = \sin x$. Then what's du? That's easy:

$$\frac{du}{dx} = (\sin x)' = \cos x$$

so $du = \cos x \, dx$. (By the way, this was in our table of differential examples that we just did.) Let's *substitute* into our integral using u and du. We get

$$\int (\sin x)^2 \cos x\, dx = \int u^2\, du.$$

I hope you're impressed. It took a lot of preliminary work to get us to this point, but substitution has transformed a messy integral into an extremely simple and easy one. Let's polish it off:

$$\int u^2\, du = \frac{1}{3}u^3 + C.$$

We can't quite stop there, because we need our final answer to be in terms of the original variable, which was x. Therefore we must remember to substitute back in for u. Here's the whole calculation, written as a unified series of steps:

$$\begin{aligned}\int (\sin x)^2 \cos x\, dx &= \int u^2\, du \\ &= \frac{1}{3}u^3 + C \\ &= \frac{1}{3}(\sin x)^3 + C.\end{aligned}$$

Sweet.

Here's one more, with a tiny extra twist. It poses no problem for the substitution technique:

$$\int e^{7x}\, dx$$

The function "inside" the integrand's exponential is $7x$, so let $u = 7x$. Then $\frac{du}{dx} = 7$, so $du = 7\, dx$. Darn! There's no 7 in the integrand to partner with the dx! Are we doomed? No, if $du = 7\, dx$, then $dx = \frac{du}{7}$. Let's go ahead with our substitution shown on the next page.

$$\begin{aligned}\int e^{7x}\,dx &= \int e^{u}\,\frac{du}{7}\\ &= \frac{1}{7}\int e^{u}\,du\\ &= \frac{1}{7}e^{u}+C\\ &= \frac{1}{7}e^{7x}+C.\end{aligned}$$

Substitution triumphs again. It works because Leibniz's differential is such a useful tool for rewriting an integral in simpler terms. With its assistance, running the chain rule backward as the anti-chain rule is like running a chain saw—you cut right through your integration problems.

From this point on, we will include the differential as a standard part of integral notation. Almost any calculus book you peek into rigorously enforces the use of the variable-identifying differential, although I carefully avoided it till now. My intention was to simplify the notation until I really needed to call on the full power of Leibniz notation, which finally happened when we got to the anti-chain rule. I confess that in my own calculus classes I never let my students write the integral sign without including the *dx* or whatever differential is appropriate for the variable we are using. I'm sure that any of them reading this book were scandalized by its absence until this chapter. I hope they feel a little better now.

23 The Best (or Worst) That You Can Be

The ups and downs of function graphs

We concluded Chapter 21 with a comment that e is the best choice for the base of an exponential or logarithmic function. It is "best" in the sense that derivatives and antiderivatives are simplest when e is the base. When it comes to mathematical functions in calculus, there are many ways in which we can judge what is "best."

In this chapter, I want to introduce you to the way that calculus lets us find when a function is best in terms of its magnitude. This process is called optimization, and we can use it to discover when a function attains its greatest possible value or its least possible value. You can easily think of situations where optimization would be useful. If you have a function describing your company's profits, you would want to know when the function is at its peak. If you have a function describing your costs, you would want to know when the function is at rock bottom. The derivative is the tool that we use to determine the optimal results.

The derivative is good for measuring rates of change. It is, after all, the slope of the tangent line. That means it can tell you whether the graph of a function is rising or falling. That seems pretty useful, particularly if we're thinking of things like the velocity of moving objects. No wonder the derivative plays an important role in physics.

What may surprise you, however, is the importance of using the derivative to find out when something is not changing, when a graph is not rising or falling. Your initial reaction might be, "That doesn't sound very interesting," but I will show you that it can be very interesting indeed.

Look at the graph in Figure 23-1. I've marked all the places where the graph is neither rising nor falling. Those are necessarily places where the tangent lines have slope $m = 0$, which we have learned is the same thing as the places where $f'(x) = 0$. (Remember, $f'(x)$ was *designed* to be the slope function for the graph of $y = f(x)$.) What can you say about the points I marked?

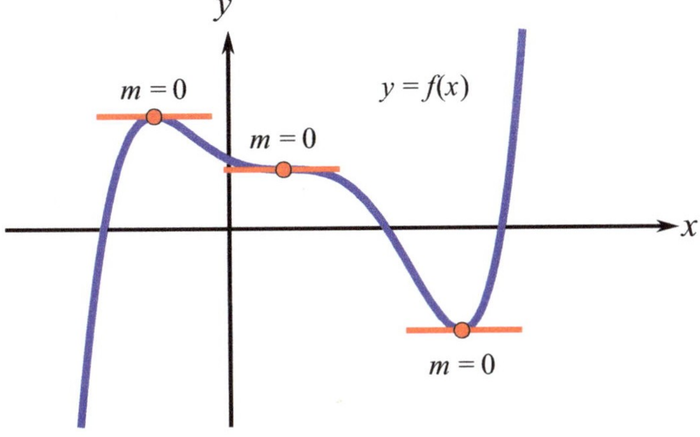

Figure 23-1.

The first point I marked (the one on the left) is a *local maximum*. All the other points in the vicinity are lower. A local maximum is the point at the top of a hill. The third point (the one on the right) is a *local minimum*. All the other points in the vicinity are higher. A local minimum is the point at the bottom of a valley.

What is going on with the point in the middle? The slope is $m = 0$, so momentarily at least the curve is flat there. It's neither a local maximum nor minimum. The middle point lets us see that the derivative (slope) can be 0 at a point without its being either a maximum or minimum. It could just be a local flat spot that is neither.

Figure 23-2 shows us a slightly different situation. This time, all three places where $f'(x) = 0$ are maximums or minimums. In fact, the maximum on the left is called a *global maximum* because it is the highest of *all* the points on the curve. (You know that this requires us to assume that nothing funny happens off the edges of the graph. I promise not to try to trick you with any examples like that.) The maximum on the right is, of course, a local maximum, and the minimum in the middle is a local minimum. You might think we should call the minimum a global minimum because it's the only one we've got, but the graph shows the curve dropping *below* the level of the minimum on both sides, so it's just local. A global minimum would have to be lower than any other point on the curve.

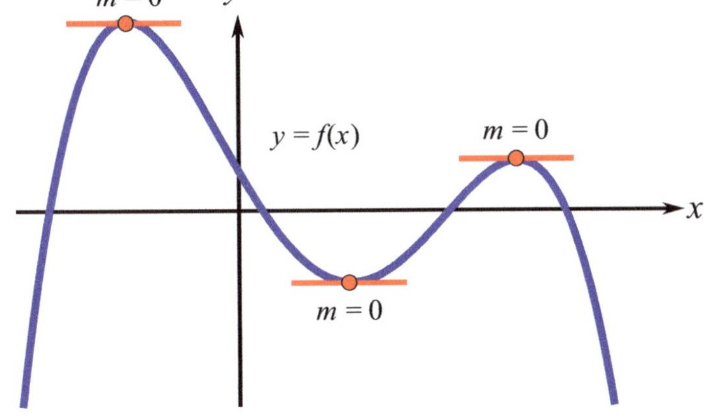

Figure 23-2.

Whether they're maximums, minimums, or neither, points where the derivative is zero are called *critical points*. In older math books, they were sometimes called stationary points in honor of the fact that their rate of change is zero, but that terminology has passed out of style.

(Don't be surprised that mathematics has fads and fashions. Like any other human endeavor, math is subject to change for change's sake. It's not stationary!)

Scaling the heights and plumbing the depths

We have now seen that critical points are the places where we should search for maximums and minimums. Let's use this observation to *optimize* a function. For example, what is the biggest possible value of the function $f(x) = 3 + 4x - x^2$? We could, of course, carefully draw a graph of it and examine the results for the highest point. (This works just fine most of the time in algebra, but we are testing out a technique that will serve us well in more challenging circumstances.) I will provide the graph soon, but for right now let's proceed without worrying about graphs. We know we need to find any critical points that $f(x) = 3 + 4x - x^2$ might have.

Critical points occur wherever $f'(x) = 0$, so let's set up that equation and solve it. The derivative is

$$\begin{aligned} f'(x) &= (3 + 4x - x^2)' \\ &= 0 + 4 - 2x \\ &= 4 - 2x. \end{aligned}$$

Take this result, plug it into the equation $f'(x) = 0$, and solve for x:

$$\begin{aligned} f'(x) &= 0 \\ 4 - 2x &= 0 \\ 4 &= 2x \\ \frac{4}{2} &= \frac{2x}{2} \\ 2 &= x. \end{aligned}$$

We call $x = 2$ our *critical number*, and we plug it into $f(x) = 3 + 4x - x^2$ to obtain the second coordinate of our critical point on the next page.

$$f(2) = 3 + 4(2) – 22 = 3 + 8 – 4 = 7.$$

Our critical point is (2, 7).

Now comes the big question: Is it a maximum? (That's what we want, because I asked for the biggest possible value of $f(x)$.) Is it a minimum? Or is it neither? Let's look very carefully at what we know. The slope function is $f'(x) = 4 – 2x$ and it equals 0 for $x = 2$: $f'(2) = 4 – 2(2) = 0$. If we plug in any number *smaller* than $x = 2$, what will happen to $f'(x)$? Try $x = 1$:

$$f'(1) = 4 – 2(1) = 4 – 2 = 2.$$

The slope turned out to be positive at $x = 1$. That means the function is rising because its rate of change is positive.

Suppose we test a number *bigger* than $x = 2$. Shall we say $x = 3$? When we try it, we get

$$f'(3) = 4 – 2(3) = 4 – 6 = –2.$$

The slope is negative at $x = 3$, so the function is *falling* because of its negative rate of change.

Let me summarize: As we move from left to right on the x axis number line, from $x = 1$ through $x = 2$ to $x = 3$, the graph of the function is first rising, then it's critical (stationary), and finally it's falling. This has to be the description of a hill. I mean, a *maximum*. Check out Figure 23-3, where I've drawn the graph of $f(x) = 3 + 4x – x^2$ and sketched in the slopes we found. The graph confirms our calculation that the biggest possible value of $f(x) = 3 + 4x – x^2$ occurs at $x = 2$ and is equal to $f(2) = 7$. The critical point (2, 7) is a maximum. What's more, it must be a global maximum, because it is the highest point of all that occur on the graph of the parabola.

Figure 23-3.

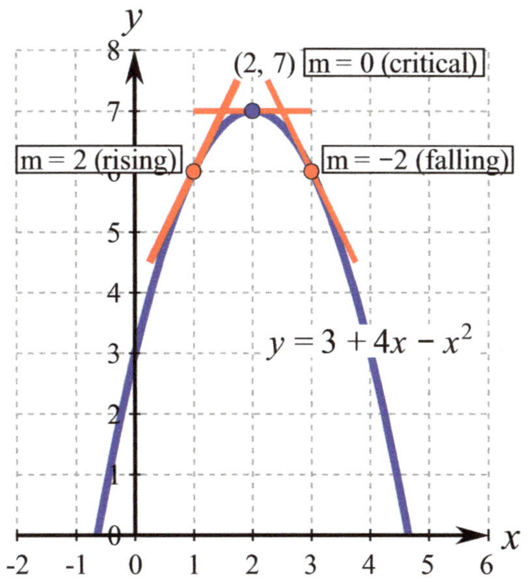

The first-derivative test

We just applied what mathematicians call the *first-derivative* test when we were trying to identify whether our critical point was a maximum, minimum, or neither. It consists of examining values of the derivative on either side of the critical number to determine what kind of critical point we have. Keep in mind that it's important to test the derivative at points fairly close to the critical point in question, because the first-derivative test is a test for *local* maxima and minima. Here's a summary table of how the first-derivative test works:

Left side	***Critical point***	***Right side***	***Conclusion***
Rising	Stationary	Falling	Maximum
Rising	Stationary	Rising	Neither
Falling	Stationary	Rising	Minimum
Falling	Stationary	Falling	Neither

Our parabola problem was an example of the first kind: rising-stationary-falling means we have a maximum.

The first-derivative test is very reliable as long as we are careful in evaluating $f'(x)$ at the various x values we choose. Keep in mind that we chose to plug in $x = 1$ and $x = 3$ because those are nice round values and they neatly bracket the critical number $x = 2$. When you have the freedom to choose your plug-in numbers, make life easy on yourself. If we had thought of it at the time, we could have used $x = 0$ as the plug-in point on the left side. It's hard to go wrong when you evaluate something at $x = 0$. (Yes, you're right. I did think of it, but I didn't use $x = 0$ because I wanted to group the values closer to $x = 2$ to make the graph nice. A teacher's life is full of compromises like this.)

Perhaps the name of the first-derivative test has given you pause. Why does it have that name, exactly? Is there a *second*-derivative test?

As a matter of fact, there is.

24 A Second-Derivative Look at Ups and Downs

Going around the bend

Let's take another look at the graph of $f(x) = 3 + 4x - x^2$. It's in Figure 24-1. When we used the first-derivative test in the previous section, we took note of the fact that the slope of the function (given by $f'(x) = 4 - 2x$) changes from positive to zero to negative. In brief, the slope is *decreasing*. What's another way to express this?

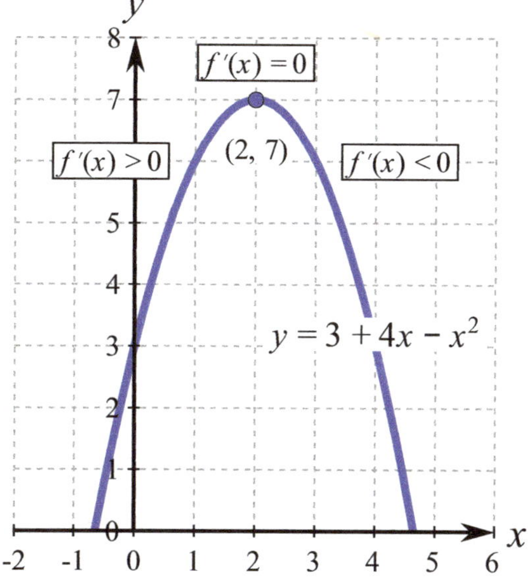

Figure 24-1.

If $f'(x)$ is decreasing, then it must have a negative rate of change. Since rate of change is given by the derivative (the *first derivative*), what will happen if we take the derivative of $f'(x)$? That's what we call a *second derivative*—the derivative of a derivative. A double-prime symbol provides the notation we need:

$$\begin{aligned} f''(x) &= (f'(x))' \\ &= (4 - 2x)' \\ &= 0 - 2 = -2. \end{aligned}$$

As we might have expected, the result is negative. Although it's rather a mouthful, we can express this by saying the rate of change of the rate of change is negative.

As we have seen, when this occurs at a critical point, the result has to be a maximum. That's because a negative second derivative means the first derivative is falling. If the first derivative is falling at a critical point (where the first derivative is zero), then it must be switching from positive to negative.

It makes sense, therefore, that a positive second derivative at a critical point implies that the critical point is a minimum. While the first-derivative test required checking a pair of points (one on each side of the critical point, both chosen fairly close to the critical point), the *second-derivative test* can be performed at just one point—the critical point itself.

If we look at something more complicated than a parabola, the second-derivative test can be a big help. For example, what are the maximum and minimum points of the function $f(x) = x^3 - 9x^2 + 24x - 11$? The critical numbers occur where $f'(x) = 0$, so we'll take care of that first:

$$\begin{aligned} f'(x) &= (x^3 - 9x^2 + 24x - 11)' \\ &= 3x^2 - 18x + 24, \end{aligned}$$

so $f'(x) = 0$ means we have to solve $3x^2 - 18x + 24 = 0$. Fortunately, it factors:

$$\begin{aligned} 3x^2 - 18x + 24 &= 0 \\ 3(x^2 - 6x + 8) &= 0 \\ 3(x - 2)(x - 4) &= 0. \end{aligned}$$

The only way this product can equal 0 is for one of its factors to equal 0. If $x - 2 = 0$, we get $x = 2$. If $x - 4 = 0$, we get $x = 4$. Our critical numbers are 2 and 4.

The second derivative is

$$\begin{aligned} f''(x) &= (f'(x))' \\ &= (3x^2 - 18x + 24)' \\ &= 6x - 18. \end{aligned}$$

When we use the second-derivative formula to test the critical numbers, we get $f''(2) = 6(2) - 18 = 12 - 18 = -6$ and $f''(4) = 6(4) - 18 = 24 - 18 = 6$. Since the second derivative is negative at $x = 2$, that must be a maximum. Since it's positive at $x = 4$, that must be a minimum. See the graph in Figure 24-2.

Figure 24-2.

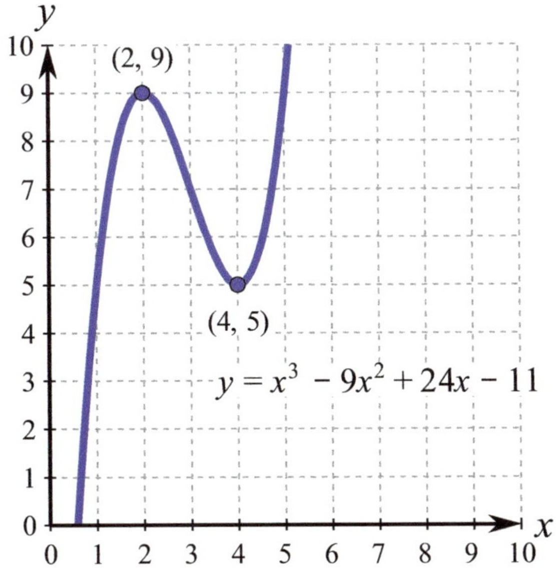

The famous farm problem

Farmer Tom has enough lumber to build a fence 80 yards long. He intends to build his fence in the shape of a rectangular enclosure. Being a frugal man, as successful farmers must be, Tom wants his rectangular corral to enclose as much area as possible. He's heard that the most economical enclosure is a square one, but Tom doesn't like to take things just on faith. He'd like to know for sure.

Fortunately, Farmer Tom knows a little calculus. He figures his fence problem would be a good one to solve with a computation of maximum area. If he can get a formula for area, he can apply the methods of calculus to find out when the area is greatest.

The area of a rectangle is given by $A = xy$, where x and y are the length and width of the rectangle. The perimeter of the rectangle (the distance around it) is just $2x + 2y$, which is going to equal 80 yards, since that's how much fence Tom can build. (See Figure 24-3.) This is what Tom knows so far:

Figure 24-3.

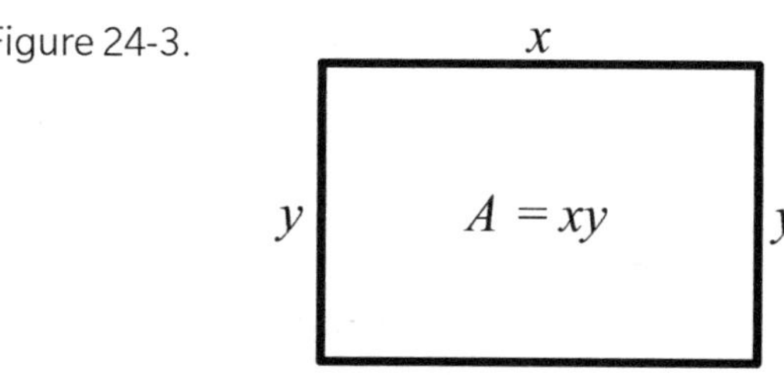

$$A = xy$$
$$2x + 2y = 80.$$

Too bad A is expressed in terms of two variables, x and y, instead of just one. That keeps us from taking its derivative and setting it equal to zero, just as we learned to do when looking for maximum and minimum values.

Let's not forget about the second equation, though. If we solve the perimeter equation for y, this is what we get:

$$\begin{aligned} 2x + 2y &= 80 \\ 2y &= 80 - 2x \\ \frac{2y}{2} &= \frac{80 - 2x}{2} \\ y &= 40 - x. \end{aligned}$$

This will do the trick, because now we can substitute for y in the area formula:

$$A = xy = x(40 - x) = 40x - x^2.$$

Farmer Tom is in business now, because the area is just a simple function of x (and only x). Let's find its derivatives:

$$\begin{aligned} A(x) &= 40x - x^2 \\ A'(x) &= 40 - 2x \\ A''(x) &= -2. \end{aligned}$$

The second derivative is useful in this case because it's always negative. Any critical number we find must be a *maximum*. Let's try to find some by setting the first derivative equal to zero:

$$\begin{aligned} A'(x) &= 0 \\ 40 - 2x &= 0 \\ 40 &= 2x \\ 20 &= x. \end{aligned}$$

The only critical number is $x = 20$. When the length is given by $x = 20$ yards, the width must be given by $y = 40 - x = 40 - 20 = 20$. The enclosure needs to be a 20 by 20 square if Farmer Tom wants to maximize its area. If we look at the graph of the area function in Figure 24-4, we can see that the maximum area is $A(20) = 20(40 - 20) = 20^2 = 400$ yd^2. Farmer Tom knows what he has to do.

Figure 24-4.

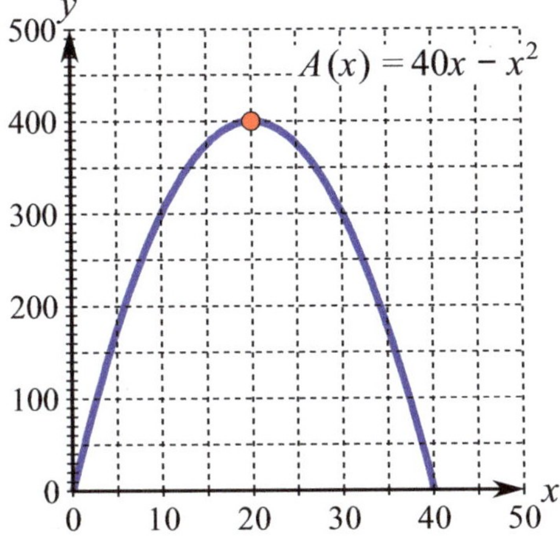

Concavity and diminishing returns

When the second derivative of a function is negative, we say that the graph of the function is *concave down*. Conversely, we say the graph is *concave up* when the second derivative is positive. Figure 24-5 shows us the four situations possible for the signs of the first and second

Figure 24-5.

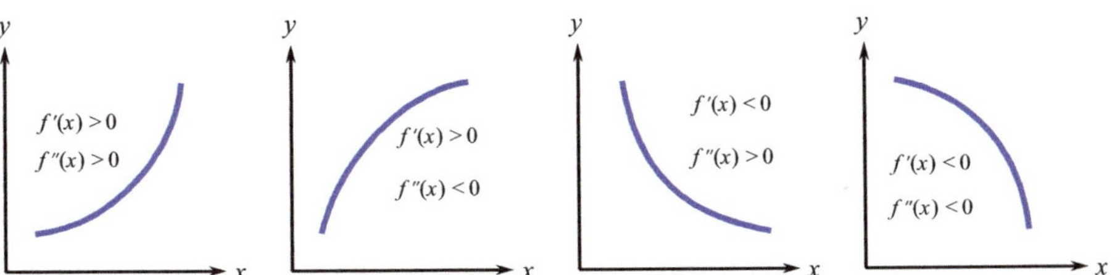

derivatives, the different combinations of rising and falling and concave up and concave down.

Check out the second graph in Figure 24-5, a rising function ($f'(x) > 0$) that is concave down ($f''(x)$). It's what we call "diminishing returns"—the situation where the curve is still increasing, but at a decreasing rate. An example could be a company spending money on advertising. At first the money creates a big increase in sales, but as the market gets saturated with advertising, the ad money generates lesser returns. Revenues keep going up, but they don't go up as fast as they initially were. Can you come up with examples for the other three cases in Figure 24-5?

I'd like to introduce you to another situation where the concavity of a curve matters quite a bit. It's the case of *logistic growth*, a model used by ecologists to represent what happens to an animal population as its environmental niche is filled to its capacity. See Figure 24-6. Suppose that a particular environmental niche can support a population of as many as M animals of a particular species. The logistic curve shows us how population growth slows down as the population nears M, the upper limit to its size. The concavity of the logistic curve changes when the population reaches $\frac{M}{2}$, half the total capacity. A scientist who collects data on an animal population in a specific locale could watch the population numbers to see when the curve stops being concave up and begins to be concave down. That point, if the scientist can determine it, will then reveal the value of $\frac{M}{2}$, after which it is easy to compute the total bearing capacity of the niche under observation.

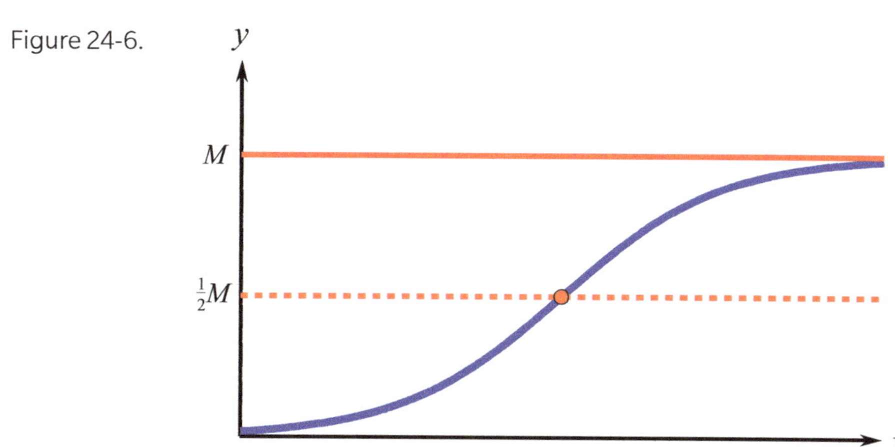

Figure 24-6.

You can try to work with the logistic curve yourself if you'd like a challenge. The function is of the form $y = \dfrac{M}{a + e^{-bx}}$, where M, a, and b are all positive constants. Find y'', the second derivative, and see if you can figure out when it equals zero. When you plug that x value back into the original function, you'll see that y equals half of M at that point. (Good luck!)

Real-life situations aren't always nice enough to give us equations to work with. Researchers are usually forced to work with raw data, and you can't really take a first or second derivative of a scattered collection of numbers. Nevertheless, I hope you can see how the notion of concavity can be useful in examining how the numbers may appear on a graph. And it was the second derivative that gave us a helpful way to talk about concavity.

The A word

There is one example of a second derivative that is so common we all know its special name. Perhaps it's already occurred to you—perhaps tipped off when I mentioned the rate of change of the rate of change—and you've been champing at the bit to blurt it out: *acceleration*. What is acceleration, after all, but the rate of change of velocity, which is itself the rate of change of distance with respect to time? In mathematical language, if $y = f(t)$ tells you where you are along some path at time t, then $y' = f'(t)$ is the velocity (the speed at which you're traveling), and $y'' = f''(t)$ is the acceleration (the rate at which your speed is changing).

If you drive a vehicle, you have probably experienced the challenge of merging into freeway traffic via an on-ramp. How do you do it? In most circumstances, a merging driver makes two assumptions without thinking too much about it: (1) The driver assumes that the other cars are traveling at constant speed, with no acceleration ($y'' = 0$). (2) The driver assumes that his or her car has constant acceleration on the on-ramp ($y'' =$ constant). With those two basic assumptions, the driver looks for a gap in the traffic where the constant speed of the traffic matches up with the constant acceleration of the merging vehicle—a place where the merging vehicle's speed and position will match the traffic gap at the end of the on-ramp. The calculus in your head is what helps you merge into traffic, although we probably don't think of it in terms of calculus (I do, sometimes, but I'm a math teacher) and we certainly don't pull out a notepad and start scratching out some derivative calculations. We do it intuitively, based on our experience. Still, this is another example of how calculus provides a model for things we experience in real life.

And, of course, we know that the basic assumptions are sometimes wrong. Like when some driver sees you trying to merge and speeds up ($y'' > 0$) to cut you off. But I won't say what I think of people like that.

I have a particular example of an acceleration problem from physics that is easily solved using calculus. In the next chapter, we will use integrals to conquer gravity.

25 We Conquer Gravity

What goes up, must be an integral

In the days when physicists were still being called natural philosophers ("lovers of knowledge of nature"), they observed that falling objects had one very special property in common: They all fell with constant acceleration. That is, the velocity of a falling object kept steadily increasing. The longer it fell, the faster it went. What's more, these early researchers figured out that the acceleration was the same for all objects and that the acceleration equals 32 feet per second per second in the old-fashioned units of the day. This acceleration is usually written as 32 ft/sec/sec or, preferably, 32 ft/sec^2.

Aristotle had thought otherwise. He said heavy objects fall faster than light objects. For example, it's obvious that feathers don't fall as fast as lead weights. The air gets in the way and fools us (and Aristotle) into thinking gravity has a harder time getting a grip on feathers than on lead weights. Commander Dave Scott of the Apollo 15 moon mission in 1971 took advantage of his time on the airless surface of the moon to do a little physics experiment. He simultaneously dropped a hammer and a feather during a live television broadcast to Earth. Viewers saw the two objects hit the lunar surface at the same time. Galileo is famously reputed to have done something similar nearly 400 years earlier from the top of the Leaning Tower of Pisa (though scholars doubt the story is true). He supposedly had to use cannon balls of different weights to avoid undue influence by air resistance, in the absence of which they hit the ground simultaneously. A feather would not have worked.

Everyone agrees, though, that—except for a few details like air resistance—falling objects all drop toward the ground with a constant acceleration of 32 ft/sec^2. We know that acceleration can be represented as a second derivative, but the second derivative of what?

Suppose we use the function $h(t)$ to represent the height above the ground at time t of a falling object. Then $h(0)$ is the initial height of the object at the moment we click the stopwatch and start keeping track of the height. Then $h(1)$ is the object's height after 1 unit of time, etc. What, then, is $h'(t)$? It must be the rate of change of height with respect to time. If the height is measured in feet and the time in seconds, then $h'(t)$ must be in terms of feet per second. That is, $h'(t)$ is velocity, the rate at which the object falls.

If we take the second derivative, $h''(t)$ will be the rate of change of the velocity, so it must be acceleration. That's exactly what we were looking for, but there's one more consideration before we can write an equation.

What is the sign of $h''(t)$? Because a falling object is rushing toward the ground, its height is decreasing at an ever greater rate. Both the velocity and acceleration of a falling object must be negative. We left out that little detail while talking about 32 ft/sec^2, which should really be expressed as a negative quantity. We can therefore write

$$h''(t) = -32 \text{ ft/sec}^2.$$

Thanks to our earlier work with integrals, we know how to "undo" a derivative. Let's slap an integral sign on both sides of the equation for $h''(t)$ and compute some antiderivatives:

$$\int h''(t) = \int (-32)$$
$$h'(t) = -32t + C_1$$

Does that look okay to you? The antiderivative of $h''(t)$ has to be $h'(t)$—we just take one prime away—and the antiderivative of a constant like –32 has to be –32t. Then there's the arbitrary constant (we mustn't forget that), which I wrote with a subscript of 1. Why?

Because we're going to do it all over again and there'll be a second constant to deal with. I'll call that one C_2, of course. Here we go:

$$\int h'(t)dt = \int (-32t + C_1)dt$$
$$h(t) = -32 \cdot \frac{1}{2}t^2 + C_1 t + C_2$$
$$h(t) = -16t^2 + C_1 t + C_2.$$

Problem solved. We have a formula for $h(t)$ and it turned out to be quadratic (second degree). But it contains two arbitrary constants that popped up because of the antiderivatives we took. What are we to do with them? It turns out that each one has a special significance. Go ahead and plug $t = 0$ into the formula for $h(t)$:

$$h(0) = -16 \cdot 0^2 + C_1 \cdot 0 + C_2$$
$$= C_2.$$

So C_2 just equals $h(0)$, but $h(0)$ is the height of the object at time $t = 0$. It's what we call the *initial height* of the dropped (or thrown) object. Now we know about C_2, so let's consider what we can figure out about C_1. The first constant appeared in the formula we found for $h'(t)$. Let's go back to $h'(t) = -32t + C_1$ and plug $t = 0$ into it:

$$h'(0) = -32 \cdot 0 + C_1$$
$$= C_1.$$

There you have it: C_1 is equal to $h'(0)$. But what does that mean? Since $h'(t)$ is the rate of change of height (a distance) with respect to time, it has units of velocity (we already knew this, right?). Thus $h'(0)$ is simply the velocity at time $t = 0$, also known as the *initial velocity*. Physicists are fond of using v_0 as the symbol for initial velocity and h_0 for initial height, where the zero subscripts are supposed to remind us that the symbols represent quantities at time zero. With this notation, our height formula becomes

$$h(t) = -16t^2 + v_0 t + h_0.$$

If we know an object's initial velocity and height, we can use the $h(t)$ formula to find its height at any time thereafter. We have completely solved the falling object problem with a little help from calculus.

Look out below!

Obadiah is a young boy on the observation deck on the 86th floor of the Empire State Building (shown in Figure 25-1), approximately 1053 ft above the ground. He's tossing a penny in his hand. Obadiah throws it up into the air with an initial velocity of 48 ft/sec. He misses it as it comes back down toward him, and he quickly loses sight of it as the penny plunges toward the ground. How long does it take for the penny to hit the ground?

Figure 25-1.

We know that $v_0 = 48$ ft/sec (because he threw it upward; it would be –48 ft/sec if he threw it downward) and $h_0 = 1053$ ft (we could quibble about whether Obadiah let go of the penny high above his head, adding to the height, but let's not). The height formula for Obadiah's penny and its fall from the Empire State Building is $h(t) = -16t^2 + 48t + 1053$. Let's plug in different values of t at one-second intervals to see how $h(t)$ changes with time:

t	$h(t)$
0	1053
1	1085
2	1085
3	1053
4	989
5	563
6	765
7	605
8	413
9	189
10	–67

Whoa! Something definitely went wrong at the 10 second mark. The penny ended up 67 feet underground! Yeah, that's likely. This just means our formula failed because we pushed it too far. (Mathematical models of reality are like that, you know. We have to be careful how much we trust them. They're just models, after all.) Nevertheless, the table tells us a lot of things about the penny's trajectory.

First, the penny rose higher into the air because Obadiah tossed it upward. After 3 seconds, however, the coin had fallen back to Obadiah's altitude at 1053 ft, at which point he failed to catch it as it went by. The penny continued to fall until at 9 seconds it was only 189 ft above the ground. Less than a second later, it hit the ground. Can we find the exact time that it hit the ground? Sure. Just set $h(t)$ equal to 0 and solve for t with the quadratic formula:

$$\begin{aligned} h(t) &= 0 \\ -16t^2 + 48t + 1053 &= 0 \\ t &= \frac{-48 \pm \sqrt{48^2 - 4(-16)(1053)}}{2(-16)} \\ &= \frac{-48 \pm \sqrt{69696}}{-32} \\ &= \frac{-48 \pm 264}{-32} = \begin{cases} -6.75 \\ 9.75 \end{cases} \end{aligned}$$

The negative answer makes no sense for our falling penny problem, so we rule that out and take 9.75 seconds as our answer. That's when the penny hits the ground.

Do we know how fast the penny was traveling when it hit? That's even easier. We compute $h'(t)$ and plug in $t = 9.75$:

$$\begin{aligned} h'(t) &= (-16t^2 + 48t + 1053)' \\ &= -16 \cdot 2t + 48 \cdot 1 + 0 \\ &= -32t + 48. \end{aligned}$$

The quadratic formula

One of the most important algebra formulas is the *quadratic formula*, which tells us that the solutions to the quadratic equation $ax^2 + bx + c = 0$ are given by

$$x = \frac{-b \pm \sqrt{b^2 - 4ac}}{2a}.$$

Therefore, $h'(9.75) = -32(9.75) + 48 = -264$ ft/sec. The penny is traveling pretty fast when its trip abruptly ends. (It turns out that 264 ft/sec is the same as 180 mi/hr.)

One last question about the penny. Do we know how high it rose before falling back? The greatest height recorded in the table is 1085 ft, which occurred both at $t = 1$ and $t = 2$. We could either surmise that something happened in between those two times or we could call on our knowledge of calculus. When do maximum values occur? At critical points. How do we find critical points? By finding out where the derivative is zero. We already know that the derivative is $h'(t) = -32t + 48$, so let's set it equal to 0 and solve:

$$h'(t) = -32t + 48 = 0$$
$$-32t = -48$$
$$t = \frac{-48}{-32} = 1.5.$$

That's between 1 and 2, all right! We see that $t = 1.5$ is a critical number. When we plug into $h(t)$ to find the height at that moment in time, we obtain $h(1.5) = 1089$ ft. That's as high as the penny ever gets. (See Figure 25-2 for a graph of $h(t)$.)

Figure 25-2.

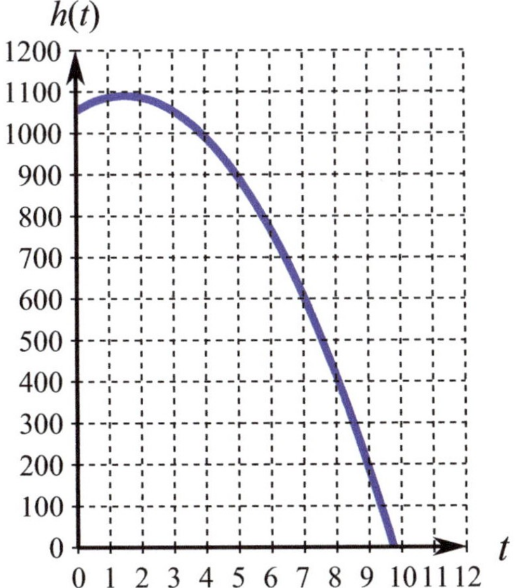

Out of this world

Now that we know all about tossing pennies, can we calculate how long it took Dave Scott's feather to fall to the surface of the moon? He dropped it from chest height, which we could take as 4.5 ft or so, and he merely let go of it, meaning there was no initial velocity imparted to it. Therefore we have $v_0 = 0$ ft/sec and $h_0 = 4.5$ ft. We would be all set to go, except that *Scott was on the moon*. The moon's gravity is much weaker, so its gravitational acceleration is not the same as the Earth's -32 ft/sec^2. Fortunately, we know that the moon's gravity is about one-sixth of the Earth's. The gravitational acceleration at the moon's surface is approximately -5.35 ft/sec^2. Recalling that this acceleration gets divided by 2 in the process of deriving (or antideriving) the height formula, we get

$$\begin{aligned} h(t) &= \frac{1}{2}(-5.35)t^2 + 0 \cdot t + 4.5 \\ &= -2.675t^2 + 4.5. \end{aligned}$$

We set this equal to 0 (ground level is always zero height) and solve for the time t:

$$\begin{aligned} -2.675t^2 + 4.5 &= 0 \\ -2.675t^2 &= -4.5 \\ t^2 &= \frac{-4.5}{-2.675} \\ t &= \sqrt{\frac{4.5}{2.675}} \approx 1.30. \end{aligned}$$

It took Dave Scott's feather about 1.3 seconds to hit the ground. The video of Commander Scott's experiment has been archived on the Web, so you might want to get out your stopwatch and check. Does it agree with our calculation of 1.3 seconds?

26 Population Boom

Doing what comes naturally

When we figured out the problem of finding the height above ground of a falling object, we started with y'', a second derivative, and worked back toward y by computing antiderivatives. The process of finding an unknown function from its derivative (or derivatives) is called *solving a differential equation*. Differential equations are important in all branches of science and engineering because they let people describe how something changes (such as in an experiment) and then work out a formula for whatever it was that changed. We had information about the rate at which height changes and used it to get a formula for height itself.

The falling object problem involved a second derivative, so we had to integrate twice. If we had a differential equation that involved only the first derivative, then we ought to be able to solve it by integrating once. (In a way, we've been solving simple differential equations every time we found an antiderivative. We just didn't call it that at the time.) We are going to solve the differential equation of *natural growth*, which says that the more you have, the more you get. If y gives the size of a population at time t, so that y' is the rate of change of that population with respect to time, then the natural growth equation says

$$y' = ky,$$

where k is some constant factor (we call it the *natural growth constant*) that depends on the particular population being studied.

There are sound biological reasons why a population grows more rapidly the bigger it gets, but I hope you'll forgive me if I skip over the details of how members of a population add to that population. This is, after all, a book about math rather than biology.

Solving the population problem

Let's rewrite the natural growth equation with Leibniz notation for the derivative:

$$\frac{dy}{dt} = ky,$$

and then turn it into a differential equation:

$$dy = ky\,dt.$$

Are you tempted to reach for the integral sign now and apply it to both sides? Hold on a moment, because that would cause a problem. The variable y is on both sides of the differential equation. That's awkward. It would be much better if we had the y's all on one side and t on the other. A little algebra does the trick. Just divide both sides by y:

$$\frac{1}{y}dy = k\,dt.$$

Now let's apply the integral sign:

$$\int \frac{1}{y}dy = \int k\,dt.$$

That's much better, right? The left side is a logarithm, and the right side is the antiderivative of a constant (with respect to t). We have it made:

$$\ln y = kt + C.$$

Our only problem is that we want to solve for y, which is stuck inside the natural logarithm function. But that's not a big difficulty. Recall that $\ln a = b$ implies that $a = e^b$, so any statement written in logarithmic language can be restated in terms of exponential language. Since $\ln y = kt + C$, we can say that

$$y = e^{kt+C}.$$

Do you remember your laws of exponents? Since $e^{x+y} = e^x e^y$, we have

$$y = e^{kt}e^C.$$

We see that y is clearly an exponential function, but what does that e^C mean? The simplest thing to do is to evaluate the equation at $t = 0$, because that's our most basic case:

$$\begin{aligned} y\big|_{t=0} &= e^{kt}e^C\big|_{t=0} \\ &= e^{k\cdot 0}e^C = e^0 \cdot e^C \\ &= 1 \cdot e^C = e^C. \end{aligned}$$

It turns out that e^C must simply equal the initial value of y, that is, the value of y at $t = 0$. We call this the *initial population* (for all of the obvious reasons). Just as we used v_0 for initial velocity and h_0 for initial height in the last chapter, we can use y_0 for the initial population. We can then rewrite the solution to the differential equation of natural growth as

$$y = y_0 e^{kt},$$

which is the classic form of the answer.

Real-world numbers

Population data from the U.S. Census Bureau shows that the mid-year world population in 2000 was approximately 6.1 billion, while in 2005 it was approximately 6.5 billion. How does our natural growth formula allow us to predict future population?

Let $y = 0$ correspond to the year 2000. Then $y_0 = 6.1$ billion and our formula becomes

$$y = 6.1e^{kt}.$$

But we still don't know what k is. Let's use our second data point to find out. If we plug in $t = 5$ for the year 2005 (it's 5 years later than 2000), the formula becomes

$$6.5 = 6.1e^{k(5)}.$$

We'll use some algebra skills to solve for k. First, divide both sides by 6.1:

$$\frac{6.5}{6.1} = e^{5k}.$$

Now rewrite this exponential statement as a logarithmic statement:

$$5k = \ln \frac{6.5}{6.1}.$$

Finally, we divide both sides by 5 and use our calculator to get a numerical result:

$$k = \frac{1}{5} \ln \frac{6.5}{6.1} \approx 0.0127.$$

Our population formula turns out to be

$$y = 6.1e^{0.0127t},$$

where t is the number of years after 2000 and y is in units of billions of people. (See Figure 26-1.)

Figure 26-1.

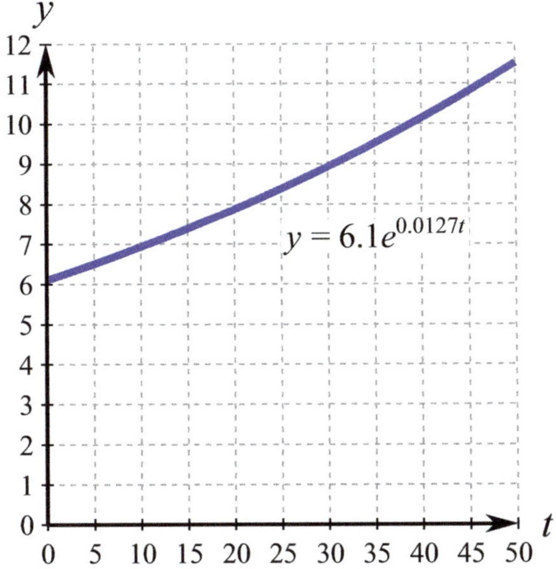

So how many people do we expect to occupy the Earth in the year 2010? 2050? For 2010, we plug in $t = 10$:

$$y = 6.1e^{0.0127(10)} \approx 6.9.$$

For 2050, we plug in $t = 50$ and obtain

$$y = 6.1e^{0.0127(50)} \approx 11.5.$$

It looks like the Earth's population will nearly double by 2050 compared to the population it had in 2000. If you have a spare room, it shouldn't be too difficult to rent it out in 2050.

Exponential curves have an interesting property: If a population doubles in size over a 50-year period, it will double over *any* 50-year period, whether it begins in 2000, 2001, or whenever. Because this period of time is a fixed quantity for any given exponential population function, mathematicians call it the *doubling time*. In our example, the result for a 50-year period was only approximately twice the initial amount. To compute a more accurate doubling time, we would have to solve the equation $e^{0.0127t} = 2$. That is, what value of t turns the exponential part of the function into the factor 2? The solution in this case is

$$t = \frac{(\ln 2)}{0.0127} \approx 54.58 \text{ years.}$$

By the way, the population estimates posted on the U.S. Census Bureau site are 6.8 billion for 2010 and 9.2 billion for 2050. Why do their numbers differ from ours? We assumed that the rate of population growth is constant, whereas it has actually been slowing down a bit. Our value of k, 0.0127, corresponds to an annual growth rate of approximately 1.27%, but the U.S. Census Bureau expects k to drop below 1% some time in the next decade or so.

It's still going to get crowded, but not quite as quickly as suggested by the natural growth equation.

27 Polynomials Forever!

The unbounded *n*th degree

Have you ever noticed that polynomials are nothing more than arithmetic with a variable tossed in? In arithmetic you take ordinary numbers and add, subtract, multiply, and divide them. If you add a variable like x to the mix, you can add x to a number, like $x + 3$, multiply by a number, like $2x$, or multiply it by itself, like x^2. All the polynomials can be built up by doing arithmetic operations on numbers and variables.

Considering how important they are in mathematics, polynomials sure are simple. For example, if someone asked you to compute the value of $3x^3 + 2x^2 - 7x + 5$ at $x = \frac{1}{2}$, you might prefer to reach for a calculator, but you know you could do it by hand, too. All it takes is arithmetic. However, you'd be right out of luck if you were asked to evaluate e^x at $x = \frac{1}{2}$. Without the calculator or a printed table of values, we'd be stuck. By comparison, evaluating the polynomial is simple.

Polynomials are simple when it comes to calculus, too. It's easy to find their derivatives (just use the power rule on each term) and just as easy to find their antiderivatives (just use the other power rule on each term). Too bad every function isn't a polynomial! Then calculus would be easy all the time.

Believe it or not, some people decided to give it a try. One of them was Brook Taylor, a contemporary of Newton, who used the tools of calculus to demonstrate how to do it. The plan was to write mathematical functions as infinite-degree polynomials. You know what the degree of a polynomial is. A first-degree polynomial can be written as $mx + b$ (what we call *linear*), and a second-degree polynomial can be written as

$ax^2 + bx + c$ (what we call *quadratic*). How would we write an infinite-degree polynomial? Like this:

$$a_0 + a_1x + a_2x^2 + a_3x^3 + a_4x^4 + ...,$$

where a_0, a_1, a_2, etc., are numbers (the coefficients of the powers of x). Assume it goes on forever.

What kinds of functions could be written this way? Can we get this to make sense somehow? Let me show you one of the most interesting cases. Let's suppose we want to write the natural exponential function as an infinite-degree polynomial. Then we would have

$$e^x = a_0 + a_1x + a_2x^2 + a_3x^3 + a_4x^4 + ...$$

But what numbers would the coefficients have to be in order to make this work? Finding out is a bit tricky, but there's one number we can figure out right away. Since $e^0 = 1$, we must have

$$\begin{aligned} e^x\big|_{x=0} &= (a_0 + a_1x + a_2x^2 + a_3x^3 + a_4x^4 + \cdots)\big|_{x=0} \\ e^0 &= a_0 + a_1 \cdot 0 + a_2 \cdot 0^2 + a_3 \cdot 0^3 + a_4 \cdot 0^4 + \cdots \\ 1 &= a_0 + 0 + 0 + 0 + 0 + \cdots \\ 1 &= a_0. \end{aligned}$$

In other words, we can now say that

$$e^x = 1 + a_1x + a_2x^2 + a_3x^3 + a_4x^4 + ...$$

Now what? Well, if we really can write the exponential function this way, then we should be able to take the derivative of both sides and get the exact same result back, since the natural exponential function is its own derivative. Here we go:

$$\begin{aligned} (e^x)' &= (1 + a_1x + a_2x^2 + a_3x^3 + a_4x^4 + \cdots)' \\ e^x &= 0 + a_1 + 2a_2x + 3a_3x^2 + 4a_4x^3 + \cdots. \end{aligned}$$

If this is valid, then e^x equals $a_1 + 2a_2x + 3a_3x^2 + 4a_4x^3 + \ldots$ as well as $1 + a_1x + a_2x^2 + a_3x^3 + a_4x^4 + \ldots$. The only way two polynomials can be equal to each other is if each pair of corresponding terms is equal. If

$$1 + a_1x + a_2x^2 + a_3x^3 + a_4x^4 + \cdots = a_1 + 2a_2x + 3a_3x^2 + 4a_4x^3 + \cdots,$$

then it must be true that the constant terms are equal, $1 = a_1$, and the first-degree terms are equal, $a_1x = 2a_2x$, so that $a_1 = 2a_2$, etc. If we make a little list, we can see the pattern clearly:

$$\begin{aligned} 1 &= a_1, \\ a_1 &= 2a_2, \\ a_2 &= 3a_3, \\ a_3 &= 4a_4, \end{aligned}$$

and so on, forever. It might be better to turn the entries in the list around and solve for the higher-degree coefficients in terms of the lower-degree ones:

$$\begin{aligned} a_1 &= 1, \\ a_2 &= \frac{1}{2}a_1 = \frac{1}{2}, \\ a_3 &= \frac{1}{3}a_2 = \frac{1}{3}\left(\frac{1}{2}\right) = \frac{1}{3 \cdot 2}, \\ a_4 &= \frac{1}{4}a_3 = \frac{1}{4}\left(\frac{1}{3 \cdot 2}\right) = \frac{1}{4 \cdot 3 \cdot 2}, \end{aligned}$$

and so on, forever.

This is a good place to introduce the factorial symbol, which you may have seen before. Do you know what 5! means? (Some of my students believe it means you're supposed to shout "*Five*!" But that's not true.) Mathematicians use it as a shorthand to represent the product $5 \cdot 4 \cdot 3 \cdot 2 \cdot 1$, so that $5! = 120$. (Look for a factorial key on your calculator. It may have $x!$ on it as a label.)

What we've discovered is that there's only one way for the exponential function to be represented as an infinite-degree polynomial, and that's if the nth coefficient is a_0 divided by n! If you look at our previous results, you'll see that this fits. The formula for a_n is given by

$$a_n = \frac{1}{n!}$$

We can now write our infinite-degree polynomial for e^x. Since $1! = 1$, $2! = 2$, $3! = 6$, $4! = 24$, and $5! = 120$, we have

$$e^x = 1 + x + \frac{1}{2}x^2 + \frac{1}{6}x^3 + \frac{1}{24}x^4 + \frac{1}{120}x^5 + \cdots.$$

What happens if we take the derivative of this expression? We're supposed to get the same thing back. Let's try it:

$$\begin{aligned}
(e^x)' &= \left(1 + x + \frac{1}{2}x^2 + \frac{1}{6}x^3 + \frac{1}{24}x^4 + \frac{1}{120}x^5 + \cdots\right)' \\
e^x &= 0 + 1 + \frac{1}{2}\cdot 2x + \frac{1}{6}\cdot 3x^2 + \frac{1}{24}\cdot 4x^3 + \frac{1}{120}\cdot 5x^4 + \cdots \\
e^x &= 1 + x + \frac{1}{2}x^2 + \frac{1}{6}x^3 + \frac{1}{24}x^4 + \cdots.
\end{aligned}$$

Yes, it worked perfectly.

The preferred mathematical name for infinite-degree polynomials is *power series*. Not all functions have power series as nice as the one for the exponential function, but power series have broad applicability in higher math.

Just a handful

Remember how, at the beginning of the chapter, I mentioned the difficulty of computing $e^{1/2}$ without the aid of a calculator or printed table of values? Let's take another look at that problem. With the aid of our power series, we see that

$$
\begin{aligned}
e^{1/2} = e^x\big|_{x=1/2} &= \left(1 + x + \frac{1}{2}x^2 + \frac{1}{6}x^3 + \frac{1}{24}x^4 + \frac{1}{120}x^5 + \cdots\right)\bigg|_{x=1/2} \\
&= 1 + \frac{1}{2} + \frac{1}{2}\left(\frac{1}{2}\right)^2 + \frac{1}{6}\left(\frac{1}{2}\right)^3 + \frac{1}{24}\left(\frac{1}{2}\right)^4 + \frac{1}{120}\left(\frac{1}{2}\right)^5 + \cdots \\
&= 1 + \frac{1}{2} + \frac{1}{8} + \frac{1}{48} + \frac{1}{384} + \frac{1}{3840} + \cdots.
\end{aligned}
$$

If we add two terms of this series, we get 1.5; three terms, 1.625; four terms is a little more work, but even by hand we would soon conclude that the four-term sum is 1.6458 (to four decimal places). What does your calculator give for $e^{1/2}$? Mine says 1.6487. Just four terms were enough to give the first three digits. That's a very good start and shows us how mathematicians equipped with power series could work out the values of functions in the days when calculators were nonexistent.

It can be helpful to look at the polynomials of different degree that are based on the power series for e^x. These are called *Taylor polynomials* in honor of Brook Taylor, whose theorem is used by mathematicians to generate power series. (Sorry, but we're not going to be able to cover Taylor's theorem in this book.) The notation $T_n(x)$ is used for the nth-degree Taylor polynomial, which is just the sum of the power series terms up through nth degree. Here are the Taylor polynomials up to degree 4, and Figure 27-1 gives their graphs along with $y = e^x$.

$$
\begin{aligned}
T_1(x) &= 1 + x \\
T_2(x) &= 1 + x + \frac{1}{2}x^2 \\
T_3(x) &= 1 + x + \frac{1}{2}x^2 + \frac{1}{6}x^3 \\
T_4(x) &= 1 + x + \frac{1}{2}x^2 + \frac{1}{6}x^3 + \frac{1}{24}x^4.
\end{aligned}
$$

You may have noticed that $T_1(x)$ produces the tangent line at $x = 0$. That's not a coincidence. It always works out that way. As you might have expected, the higher-degree Taylor polynomials are better approximations for the exponential function than the lower-degree polynomials. More accuracy takes more terms and more work, of

Figure 27-1.

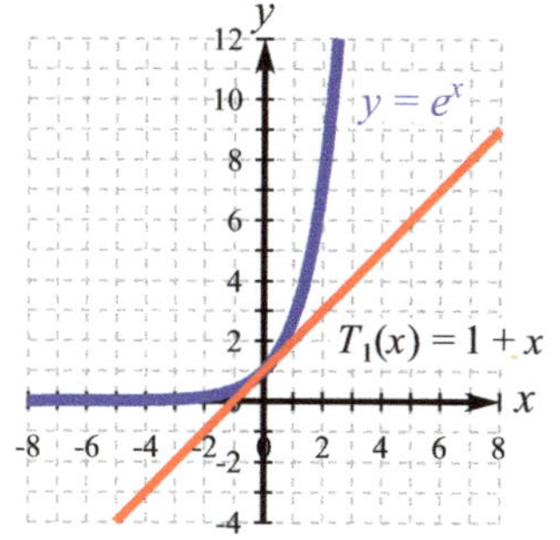

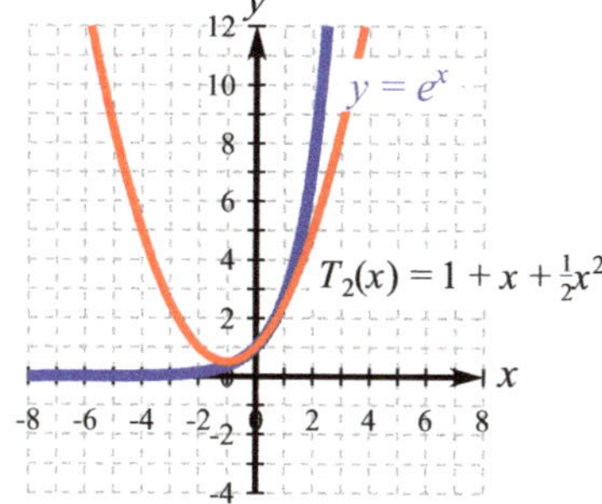

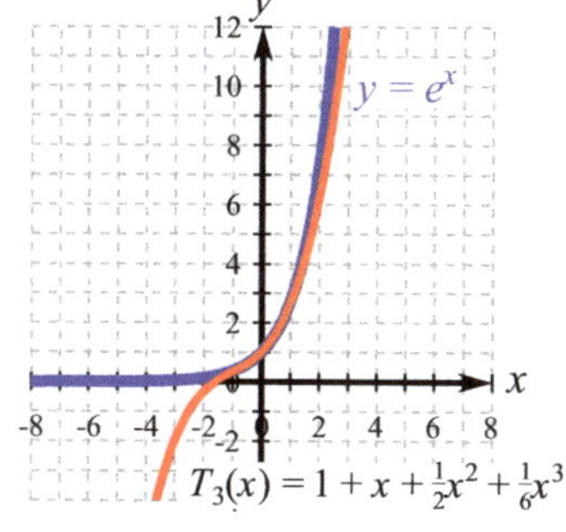

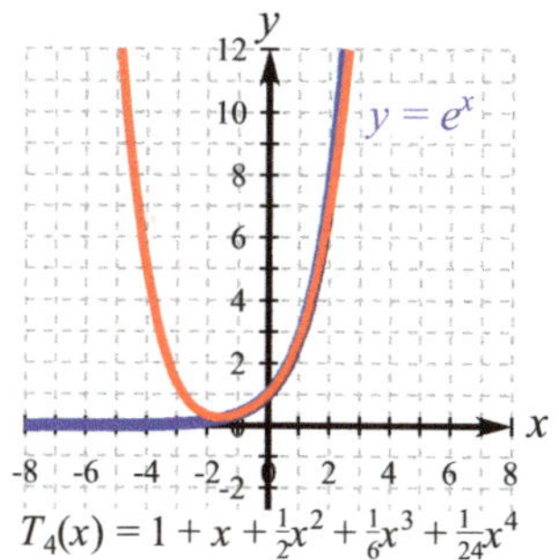

course. However, even the fourth-degree Taylor polynomial gives a good approximation only so far. As we see from its graph, $T_4(x)$ is a good match for e^x only as long as x is not too far from 0.

In case you'd like to experiment a little more with power series, I leave you with the series for our two primary trigonometric functions, sine and cosine. Try taking the derivative of the power series for sine to see if you get (as expected) the series for cosine. If you take the derivative of the power series for cosine, you should expect to obtain the *negative* of the series for sine, in keeping with the rule $(\cos x)' = -\sin x$. Check it out!

$$\sin x = x - \frac{1}{3!}x^3 + \frac{1}{5!}x^5 - \frac{1}{7!}x^7 + \cdots$$
$$\cos x = 1 - \frac{1}{2!}x^2 + \frac{1}{4!}x^4 - \frac{1}{6!}x^6 + \cdots$$

Postscript

You may remember that a polynomial is called an *algebraic function* while trigonometric and exponential functions are *transcendental*. So how can an exponential function (transcendental) equal a power series that looks like an infinite-degree polynomial? Well, all bets are off when infinity gets involved. Any finite-degree polynomial has to be algebraic, but power series go forever and can definitely represent transcendental functions, as we just saw in the specific case of e^x. Complications like this can make calculus challenging when you try to delve into the details.

28 So Very Closely Related

The conjoined twins of calculus

We're not done with the chain rule yet. One of its more potent applications is *related rates*. That's what we call the problem of figuring out how the change in one quantity might affect the change in some other quantity.

For example, suppose you're riding a Ferris wheel. It spins at a certain rate, which we could express in terms of radians/sec, while the passengers move about the circumference at a certain speed, which we could express in terms of ft/sec. How do we relate these rates to each other?

To solve the Ferris wheel problem, let's begin by recalling that the circumference of a circle of radius r is $2\pi r$. That's the full circumference, which corresponds to an angle of radian measure 2π (the angle that corresponds to going all the way around the circle). What if we have a smaller angle θ, which corresponds to some distance s along the circumference? (See Figure 28-1.) Then we have a simple proportion, because θ will be the same fraction of 2π as s is of $2\pi r$:

$$\frac{\theta}{2\pi} = \frac{s}{2\pi r}.$$

If we cross-multiply and cancel 2π from both sides, we discover that $s = r\theta$. Supposing that r is a constant (as indeed it is for our Ferris wheel), this is what happens when we take the derivative of both sides of the equation with respect to t as shown on the next page.

Figure 28-1.

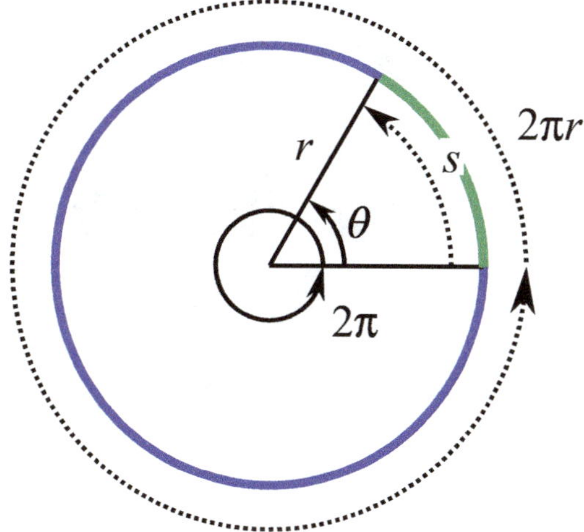

$$\frac{d}{dt}(s) = \frac{d}{dt}(r\theta)$$
$$\frac{ds}{dt} = r\frac{d\theta}{dt}.$$

To apply this formula, we need to have a specific Ferris wheel in mind. The original carnival ride to bear the name Ferris wheel was built by George Ferris, Jr., for the World's Columbian Exposition, held in Chicago in 1893. (See Figure 28-2.) Its radius was 125 ft and it turned at a rate of 3 revolutions per hour. Since each revolution is 2π radians, that turns into an angular speed of 6π radians/hour. That is our value for $\frac{d\theta}{dt}$, the angular speed. Using $r = 125$ ft, we have

$$\frac{ds}{dt} = r\frac{d\theta}{dt}$$
$$\frac{ds}{dt} = (125\text{ ft})(6\pi\text{ radians/hr})$$
$$= 750\pi\text{ ft/hr.}$$

Figure 28-2.

If we substitute a decimal approximation for π, we end up with about 2356 ft/hr. Converting this to miles per hour gives us 0.45 mi/hr. If you were riding on George Ferris's big wheel, you would have been traveling at less than half a mile per hour.

Our example can readily be used in any other case of circular motion. A more mundane problem than the Ferris wheel is a compact disc. A CD typically spins at a rate of 3.5 revolutions per second. Since each revolution is 2π radians, that turns into an angular speed of 7π radians/sec. That is our value for $\frac{d\theta}{dt}$. The radius of a CD is 6 cm. Using this value for r, we have

$$\frac{ds}{dt} = r\frac{d\theta}{dt}$$
$$\frac{ds}{dt} = (6\text{ cm})(7\pi\text{ radians/ sec})$$
$$= 42\pi\text{ cm/ sec.}$$

The vanishing radian

Why did the product (ft)(radians/hr) reduce to just ft/hr? Where did the radians go? Since the equation $s = r\theta$ can be rewritten as $\theta = s/r$, we see that the radian measure of an angle is the ratio between two lengths—a segment of the circle's circumference and its radius. Since both lengths will have the same units of measure (feet, cm, or whatever), they'll cancel and θ will turn out to be a "dimensionless" number. We just label it as "radians" to remember we're using it for angle measure.

Using a decimal approximation for π, we get about 132 cm/sec, or 1.32 m/sec. That's how fast points on the circumference of the CD are moving when the disc is spinning 3.5 revolutions per second.

Going with the flow

The related rates technique can be applied to any two quantities that are connected by an equation. Consider the volume of a cone. That's given by

$$V = \frac{1}{3}\pi r^2 h.$$

I'll use this to show you how the rate at which water drains from a conical tank is related to the changing height (depth) of the water remaining within. For my particular example, let's assume that the cone has both its height and its radius equal to 8 ft. If the water is pouring out of the tank at the rate of 10 cubic feet per second, how rapidly is the water level falling inside the tank?

Figure 28-3 shows the general situation. Because of the shape of the tank (height = radius), the amount of water in the tank at any given time will be equal to

$$\begin{aligned} V &= \frac{1}{3}\pi r^2 h \\ &= \frac{1}{3}\pi (h)^2 h \\ &= \frac{1}{3}\pi h^3. \end{aligned}$$

Using the chain rule to find the derivative of both sides of the volume equation with respect to time t, we get the result on the next page.

Figure 28-3.

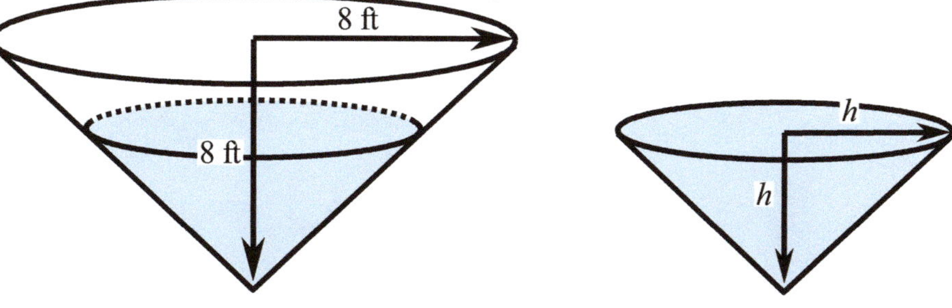

$$\frac{d}{dt}(V) = \frac{d}{dt}\left(\frac{1}{3}\pi h^3\right)$$
$$\frac{dV}{dt} = \frac{1}{3}\pi \cdot 3h^2 \frac{dh}{dt}$$
$$\frac{dV}{dt} = \pi h^2 \frac{dh}{dt}.$$

We can solve this for $\frac{dh}{dt}$ by dividing both sides by πh^2:

$$\frac{dh}{dt} = \frac{1}{\pi h^2}\frac{dV}{dt}.$$

Since we already know that $\frac{dV}{dt}$ = 10 ft³/sec (that's how fast the water is running out), all we need is a value for h and then the value of $\frac{dh}{dt}$ can be computed. Let's find out how fast the water level is falling when the depth is 5 ft. Our result is

$$\left.\frac{dh}{dt}\right|_{h=5} = \left.\frac{1}{\pi h^2}\frac{dV}{dt}\right|_{h=5}$$
$$= \frac{1}{\pi \cdot (5\text{ ft})^2} \cdot 10\text{ ft}^3/\text{sec}$$
$$= \frac{10\text{ ft}^3/\text{sec}}{25\pi\text{ ft}^2}$$
$$= \frac{2}{5\pi}\text{ ft/sec}.$$

With a little help from a calculator, we find that the water level is falling at an approximate rate of 0.127 ft/sec. If we switch to inches, it becomes approximately 1.5 in/sec. That's how fast the water is dropping when the level in the tank is down to 5 ft. Thanks to our related rates equation, we could compute this for any value of h we want, always assuming (for now, anyway) that the water is pouring out of the tank at 10 ft³/sec. A table of results appears on the next page.

Water depth h (ft)	Depth rate of change (ft/sec)
8	0.050
7	0.065
6	0.088
5	0.127
4	0.199
3	0.354
2	0.796
1	3.183

Not too surprisingly, we see that the depth falls very slowly when the tank is full, but speeds up as the water gets shallow. That's because the tank is conical and has less capacity at its pointed end. By the time the water level is down to 1 foot, the level is falling very rapidly indeed: more than 3 ft/sec.

29 What Are the Chances?

We can probably do this

Most introductory statistics courses teach you about things like the probability of getting 7 when you roll a pair of dice. Or the chances of getting two heads out of three tosses of a coin. This kind of probability is called *discrete* probability because it comes in lumps. When you roll a die, you have only six possible outcomes. If you roll a 2, then you have a 2. There are no halfway results like 2.5.

Figure 29-1.

But sometimes you have probabilities that aren't lumpy. The probability is smoothed out over more possibilities. When you flick the arrow on a spinner like the one in Figure 29-1 you also have six possible outcomes, but the situation is more complicated than rolling a six-sided die. The spinner is an example of *continuous* probability because the arrow can stop at any intermediate point—not just at a specific number. In fact, I drew the figure so that the arrow is pointing close to halfway between 1 and 2. Is it 1.5? No, not quite. Maybe 1.4. See? There's a lot of room for complications.

Of course, if we were actually using the spinner to play a game, we could just pretend it's a replacement for a die. Spin it, read the nearest whole number, and move your token on the game board—just like using dice. *However*, if you needed to work out more complicated probabilities like, say, the chances of spinning a number between 1.35 and 3.75, what would you do? What mathematical method could we apply to figure that out? Well, what else? Calculus.

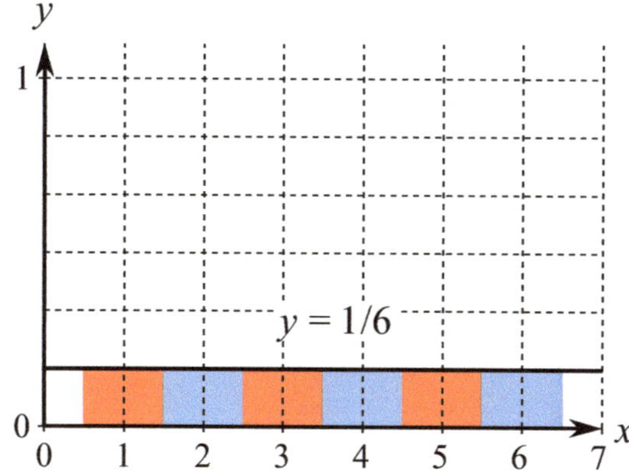

Figure 29-2.

Figure 29-2 shows a different view of the spinner. It doesn't look much like a spinner anymore. Instead it's an odd sort of graph, where the colored sections from the spinner have been marked out on the x axis.

And what's the deal with $y = \frac{1}{6}$?

It's very simple: We're trying to spread out the probability evenly over all the possible outcomes. The colored areas go from $x = 0.5$ to $x = 6.5$, as we can see from the graph. What do we get if we integrate $y = \frac{1}{6}$ from 0.5 to 6.5? We get

$$\begin{aligned}\int_{0.5}^{6.5} \frac{1}{6}dx &= \frac{1}{6}x\Big|_{0.5}^{6.5} \\ &= \frac{1}{6}(6.5 - 0.5) \\ &= \frac{1}{6} \cdot 6 = 1.\end{aligned}$$

In probability, the number 1 is the number of certainty, like 100%. The graph of $y = \frac{1}{6}$ from 0.5 to 6.5 is what mathematicians call a *probability density function* or *pdf* because its integral comes out to be 1. Any positive function with an integral of 1 can be used as a pdf. That's why we had to use $y = \frac{1}{6}$ in this case. We can apply it to find other probabilities with the following formula, where $\Pr[a \le x \le b]$ means "the probability of x ending up between a and b":

$$\Pr[a \le x \le b] = \int_a^b f(x)dx.$$

Earlier I casually wondered about the chances of spinning a number between 1.35 and 3.75. Using $y = \frac{1}{6}$ as our pdf for the spinner, we have

$$\begin{aligned}\Pr[1.35 \le x \le 3.75] &= \int_{1.35}^{3.75} \frac{1}{6}dx \\ &= \frac{1}{6}x\Big|_{1.35}^{3.75} \\ &= \frac{1}{6}(3.75 - 1.35) \\ &= \frac{1}{6}(2.40) = 0.40.\end{aligned}$$

The probability is 0.40. In other words, there's a 40% chance of getting the spinner's arrow to stop between 1.35 and 3.75.

Here are some other examples of pdf's. If you integrate the pdf over the given interval, the result has to be 1. Check out the accompanying graphs in Figure 29-3.

$$f(x) = \frac{1}{8}x, \quad [0,4]$$
$$g(x) = \frac{1}{2}\sin x, \quad [0,\pi]$$
$$h(x) = \frac{3}{32}(-x^2 + 6x - 5), \quad [1,5].$$

You can check your integration skills by verifying that each pdf encloses an area equal to 1, just as required.

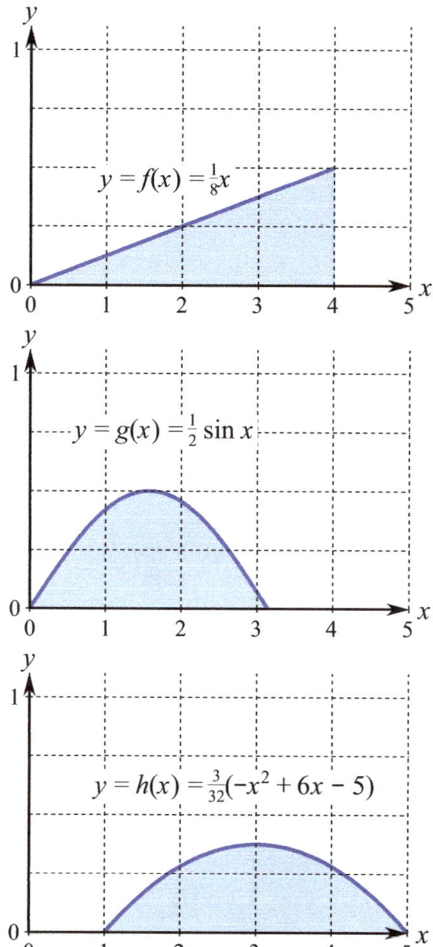

Figure 29-3.

Situation normal

The most famous pdf in mathematical probability has many names. It's variously known as the normal curve, the Gaussian distribution, or the bell curve. When teachers talk about "grading on the curve," the bell curve is what they're talking about. The actual pdf is given by the formula

$$f(x) = \frac{1}{\sqrt{2\pi}}e^{-x^2/2},$$

where x can go *forever* in both directions (or, as mathematicians like to say, from $-\infty$ to ∞). In theory, at least, the bell curve x can have any value on the whole number line. The bell curve's graph is illustrated in Figure 29-4. Although the bell curve goes forever in both directions, you can see from the figure that it drops close to zero very quickly. The probability of getting a result from the bell curve between $x = 0$ and $x = 1$ is supposed to be

$$\Pr[0 \le x \le 1] = \int_0^1 \frac{1}{\sqrt{2\pi}}e^{-x^2/2}dx,$$

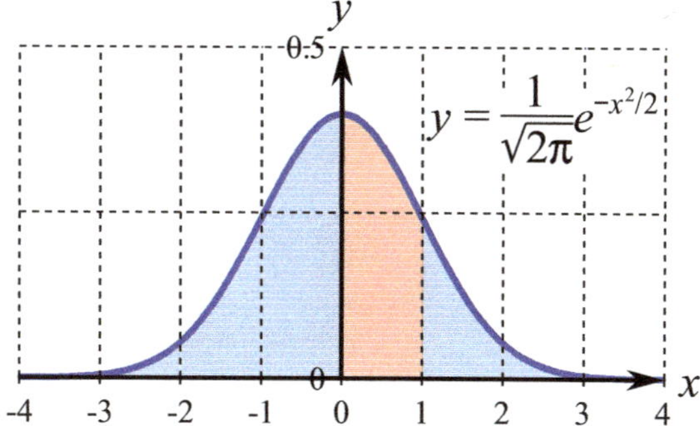

Figure 29-4.

but how are we supposed to evaluate this? As it turns out, despite the FTC, there is no convenient antiderivative for this integrand. Of course, for an important pdf like the bell curve, there are helpful reference materials that contain detailed tables of values (see the inside back cover of almost any statistics textbook). But someone must have worked it out once in order to fill in those tables. How?

Let's draw on what we learned about power series. We know one for e^x:

$$e^x = 1 + x + \frac{1}{2}x^2 + \frac{1}{6}x^3 + \frac{1}{24}x^4 + \frac{1}{120}x^5 + \cdots.$$

If we replace each occurrence of x with $-\frac{x^2}{2}$ and simplify, we get

$$\begin{aligned} e^{-x^2/2} &= 1 + \left(-\frac{x^2}{2}\right) + \frac{(-x^2/2)^2}{2} + \frac{(-x^2/2)^3}{6} + \frac{(-x^2/2)^4}{24} + \frac{(-x^2/2)^5}{120} + \cdots \\ &= 1 - \frac{x^2}{2} + \frac{x^4}{8} - \frac{x^6}{48} + \frac{x^8}{384} - \frac{x^{10}}{3840} + \cdots. \end{aligned}$$

The result is a little complicated, but easy to integrate:

$$\begin{aligned} \int_0^1 \frac{1}{\sqrt{2\pi}} e^{-x^2/2}\,dx &= \frac{1}{\sqrt{2\pi}} \int_0^1 \left(1 - \frac{x^2}{2} + \frac{x^4}{8} - \frac{x^6}{48} + \frac{x^8}{384} - \frac{x^{10}}{3840} + \cdots\right)dx \\ &= \frac{1}{\sqrt{2\pi}} \left(x - \frac{x^3}{2\cdot 3} + \frac{x^5}{8\cdot 5} - \frac{x^7}{48\cdot 7} + \frac{x^9}{384\cdot 9} - \frac{x^{11}}{3840\cdot 11} + \cdots\right)\Bigg|_0^1 \\ &= \frac{1}{\sqrt{2\pi}} \left(x - \frac{x^3}{6} + \frac{x^5}{40} - \frac{x^7}{336} + \frac{x^9}{3456} - \frac{x^{11}}{42240} + \cdots\right)\Bigg|_0^1 \\ &= \frac{1}{\sqrt{2\pi}} \left(1 - \frac{1}{6} + \frac{1}{40} - \frac{1}{336} + \frac{1}{3456} - \frac{1}{42240} + \cdots\right) \\ &\approx \frac{1}{\sqrt{2\pi}}(0.8556) \approx 0.3413. \end{aligned}$$

Are you out of breath? Look at Figure 29-5 on the next page, which shows a normal distribution table from the inside back cover of an elementary statistics textbook. I've blown up the portion of the table where you can see the entry next to $x = 1.0$, the upper limit of our integral. As you can see, we nailed the result, 0.3413.

With a little bit of patience and the tools of calculus, we could have filled in the entire table, just as mathematicians once had to do by hand. Lucky for us, those days are over!

Figure 29-5.

TABLE A-2

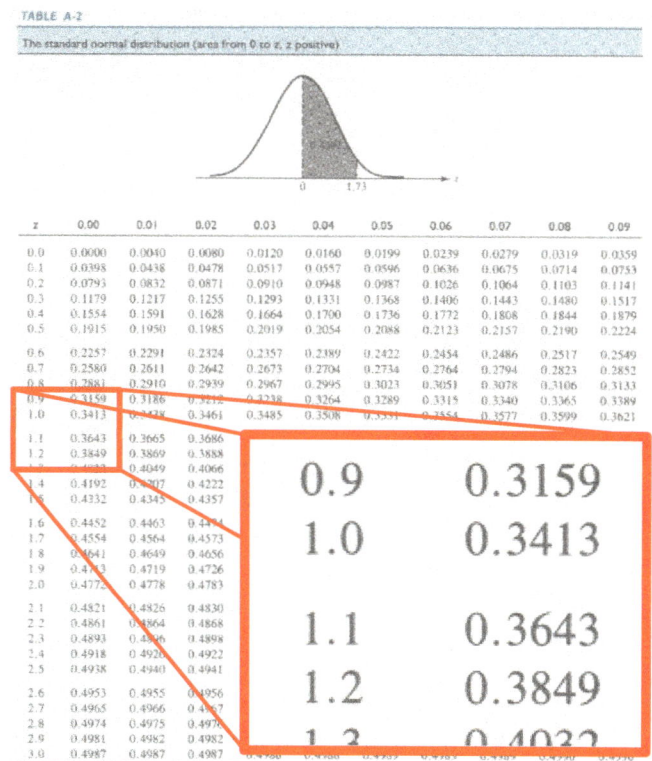

The standard normal distribution (area from 0 to z, z positive)

z	0.00	0.01	0.02	0.03	0.04	0.05	0.06	0.07	0.08	0.09
0.0	0.0000	0.0040	0.0080	0.0120	0.0160	0.0199	0.0239	0.0279	0.0319	0.0359
0.1	0.0398	0.0438	0.0478	0.0517	0.0557	0.0596	0.0636	0.0675	0.0714	0.0753
0.2	0.0793	0.0832	0.0871	0.0910	0.0948	0.0987	0.1026	0.1064	0.1103	0.1141
0.3	0.1179	0.1217	0.1255	0.1293	0.1331	0.1368	0.1406	0.1443	0.1480	0.1517
0.4	0.1554	0.1591	0.1628	0.1664	0.1700	0.1736	0.1772	0.1808	0.1844	0.1879
0.5	0.1915	0.1950	0.1985	0.2019	0.2054	0.2088	0.2123	0.2157	0.2190	0.2224
0.6	0.2257	0.2291	0.2324	0.2357	0.2389	0.2422	0.2454	0.2486	0.2517	0.2549
0.7	0.2580	0.2611	0.2642	0.2673	0.2704	0.2734	0.2764	0.2794	0.2823	0.2852
0.8	0.2881	0.2910	0.2939	0.2967	0.2995	0.3023	0.3051	0.3078	0.3106	0.3133
0.9	0.3159	0.3186	0.3212	0.3238	0.3264	0.3289	0.3315	0.3340	0.3365	0.3389
1.0	0.3413	0.3438	0.3461	0.3485	0.3508	[illegible]	0.3554	0.3577	0.3599	0.3621
1.1	0.3643	0.3665	0.3686	[illegible]	[illegible]	[illegible]	[illegible]	[illegible]	[illegible]	[illegible]
1.2	0.3849	0.3869	0.3888	[illegible]	[illegible]	[illegible]	[illegible]	[illegible]	[illegible]	[illegible]
[illegible]	[illegible]	0.4049	0.4066	[illegible]	[illegible]	[illegible]	[illegible]	[illegible]	[illegible]	[illegible]
1.4	0.4192	[illegible]	0.4222	[illegible]	[illegible]	[illegible]	[illegible]	[illegible]	[illegible]	[illegible]
[illegible]	0.4332	0.4345	0.4357	[illegible]	[illegible]	[illegible]	[illegible]	[illegible]	[illegible]	[illegible]
1.6	0.4452	0.4463	[illegible]	[illegible]	[illegible]	[illegible]	[illegible]	[illegible]	[illegible]	[illegible]
1.7	0.4554	0.4564	0.4573	[illegible]	[illegible]	[illegible]	[illegible]	[illegible]	[illegible]	[illegible]
1.8	0.4641	0.4649	0.4656	[illegible]	[illegible]	[illegible]	[illegible]	[illegible]	[illegible]	[illegible]
1.9	[illegible]	0.4719	0.4726	[illegible]	[illegible]	[illegible]	[illegible]	[illegible]	[illegible]	[illegible]
2.0	0.4772	0.4778	0.4783	[illegible]	[illegible]	[illegible]	[illegible]	[illegible]	[illegible]	[illegible]
2.1	0.4821	0.4826	0.4830	[illegible]	[illegible]	[illegible]	[illegible]	[illegible]	[illegible]	[illegible]
2.2	0.4861	0.4864	0.4868	[illegible]	[illegible]	[illegible]	[illegible]	[illegible]	[illegible]	[illegible]
2.3	0.4893	[illegible]	0.4898	[illegible]	[illegible]	[illegible]	[illegible]	[illegible]	[illegible]	[illegible]
2.4	0.4918	[illegible]	0.4922	[illegible]	[illegible]	[illegible]	[illegible]	[illegible]	[illegible]	[illegible]
2.5	0.4938	0.4940	0.4941	[illegible]	[illegible]	[illegible]	[illegible]	[illegible]	[illegible]	[illegible]
2.6	0.4953	0.4955	0.4956	[illegible]	[illegible]	[illegible]	[illegible]	[illegible]	[illegible]	[illegible]
2.7	0.4965	0.4966	[illegible]	[illegible]	[illegible]	[illegible]	[illegible]	[illegible]	[illegible]	[illegible]
2.8	0.4974	0.4975	[illegible]	[illegible]	[illegible]	[illegible]	[illegible]	[illegible]	[illegible]	[illegible]
2.9	0.4981	0.4982	0.4982	[illegible]	[illegible]	[illegible]	[illegible]	[illegible]	[illegible]	[illegible]
3.0	0.4987	0.4987	0.4987	[illegible]	[illegible]	[illegible]	[illegible]	[illegible]	[illegible]	[illegible]

0.9	0.3159
1.0	0.3413
1.1	0.3643
1.2	0.3849

30 Once Is Not Enough

Integral of integrals

In our stroll through calculus, we've seen that definite integrals can be used to compute distances, areas, volumes, and probabilities. These examples validated the point I made in the early chapters that the integral is a powerful and flexible tool. It doesn't just compute the area under a curve (as you've known since Chapter 3). In this chapter, I will take it all one step further by combining the integrals themselves.

Back in Chapter 8, I showed you that we could find volumes by integrating area functions. We worked in particular with solids of revolution because those objects had simple cross-sectional areas—always πr^2 for some suitable radius r. But what if we could handle more complicated cross-sectional areas? That is, we know that

$$\text{Volume} = \int_a^b A(x)dx$$

if $A(x)$ is the cross-sectional area function of a solid object (as in Figure 30-1). If $A(x)$ isn't a nice a simple function, perhaps we could find it with another integral. After all, if $A(x)$ is the area under the curve given by some function $f(y)$ for $c \le y \le d$, then

$$A(x) = \int_c^d f(y)dy.$$

When we plug the integral for $A(x)$ into the volume integral, we get a double integral:

$$\text{Volume} = \int_a^b \left(\int_c^d f(y)dy \right) dx.$$

Figure 30-1.

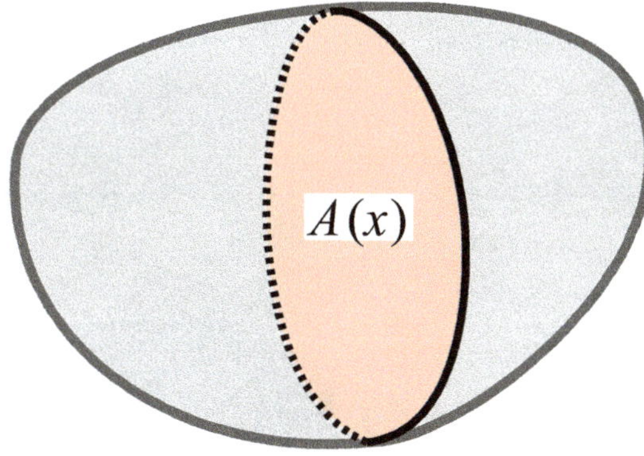

I'm going to show you how we can embed a solid object in a three-dimensional coordinate system and use that system with a double integral to find the volume of the solid object.

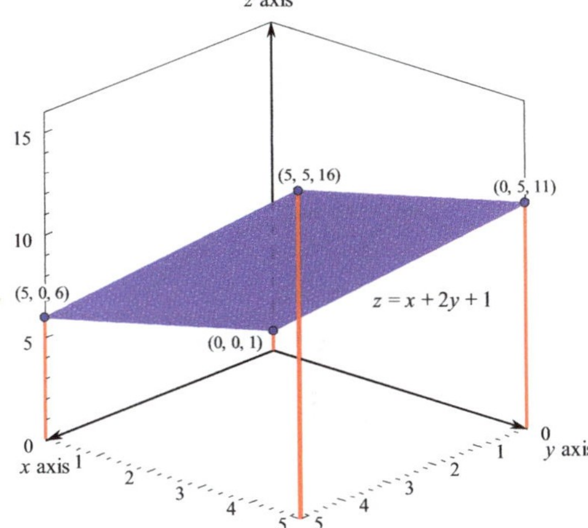

Figure 30-2.

Not lost in space

The three-dimensional version of the familiar *xy* Cartesian coordinate system is the *xyz* system. No surprise there. The *x* axis and *y* axis are perpendicular to each other, as before, but now they lie flat, forming a horizontal plane. There is also a third axis, a *z* axis, that is at right angles to the plane of the *xy* system. In Figure 30-2, you can see how I use the *xyz* coordinate system to graph the equation $z = x + 2y + 1$. Here's a short table of values that corresponds to the points that I plotted and labeled in the three-dimensional graph. For example, if I choose $x = 5$ and $y = 0$, I get $z = x + 2y + 1 = 5 + 2(0) + 1 = 6$, recorded in the table as the three-dimensional point (5, 0, 6).

x	y	z	(x, y, z)
0	0	1	(0, 0, 1)
5	0	6	(5, 0, 6)
5	5	16	(5, 5, 16)
0	5	10	(0, 5, 11)

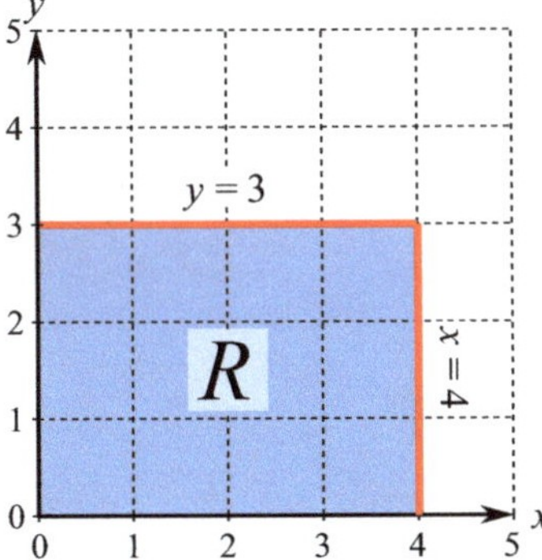

Figure 30-3.

I would like to find the volume of a solid that has a rectangular base in the *xy* plane (that's the horizontal plane at the bottom of the *xyz* coordinate system) and whose top is the plane shown in Figure 30-2—the plane whose equation is $z = x + 2y + 1$. For my rectangular base, which I'll call *R* for short, let's use our old buddy from Chapter 1. As you can see in Figure 30-3, *R* consists of all the points where $0 \leq x \leq 4$ and $0 \leq y \leq 3$.

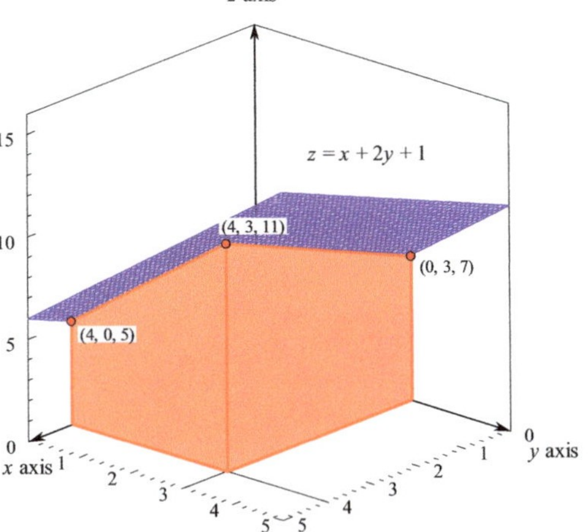

Figure 30-4.

Using *R* as the base of my solid object, let me redraw Figure 30-2, showing you the solid object I have in mind. As you can see in Figure 30-4, the top corners are marked with their coordinates: (0, 0, 1), (4, 0, 5), (4, 3, 11), and (0, 3, 7). We can't see *R*, of course, because it's underneath the solid object and not visible from this point of view. The question now is how to set up a calculation that will give us the volume of the solid.

I pretty much gave away the secret when I mentioned double integrals at the beginning of this chapter. Here's the plan of attack: I propose to "slice" through the solid object with a collection of vertical planes. Each one will produce a cross-section of the solid. We're going to find the area of each of these cross-sections, using an integral to do the job. Once we do that, we're going to have a cross-sectional area function on our hands. As we know, integrate that and you get volume. Our goal is in sight.

Check out Figure 30-5, which illustrates how we're going to chop cross-sections out of our solid. Three cross-sections are shown. In the first part of Figure 30-5, it's just $A(0)$, the cross-section we get when slicing though the solid with a vertical plane that coincides with the y and z axes. I call it $A(0)$ because $x = 0$ is the x coordinate of the slice. The third part of Figure 30-5, at the bottom, shows the cross-section $A(4)$. This cross-section is parallel to $A(0)$, but it's located where the x coordinate is now 4. Both of these cross-sections are simple vertical trapezoids whose areas are easy to find. (You might want to check that $A(0) = 12$ and $A(4) = 24$.)

The real pay-off, of course, comes with the middle illustration in Figure 30-5. It shows $A(x)$, the slice we get for some arbitrary value of x. I had to draw it somewhere, of course—really close to $x = 2.7$ in this case—but think of it as just being any old place between $x = 0$ and $x = 4$. Can we compute the value of $A(x)$, even if we don't know the actual value of x? We sure can!

Check out Figure 30-6 on the next page. It shows $A(x)$ embedded in a coordinate system. Look at the axis labels. The horizontal axis is y and the vertical axis is z, just as in the previous graphs. It's the graph of the slice that gives us $A(x)$. What's x? Well, we're not saying. We're going to pretend it's any old value between 0 and 4 (sound familiar?

Figure 30-5.

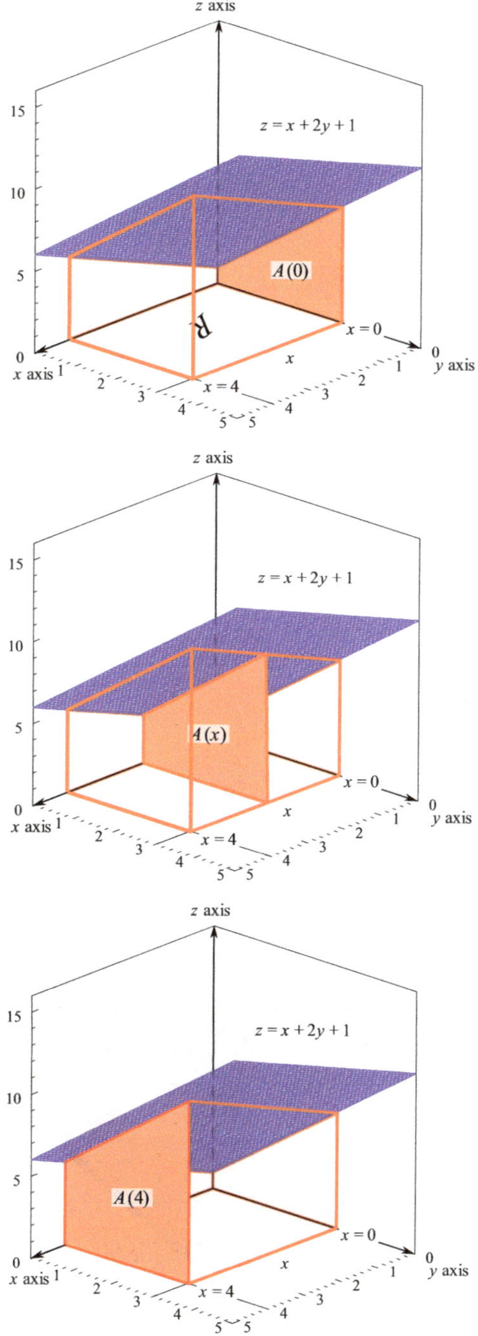

I'm just repeating myself), and for the duration of this calculation, *x is a constant*. Got that? We've picked a value and we're sticking with it. At least for now.

Figure 30-6.

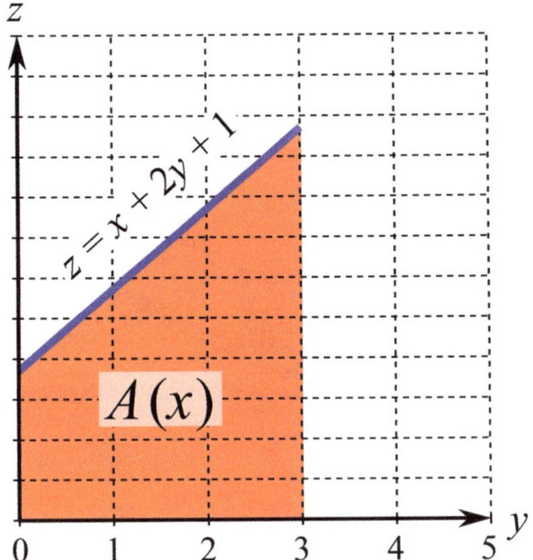

Finding area is a cinch with calculus. We know that $A(x)$ is simply the area under the curve (the line, really) that is shown in the figure. What's its equation? Just $z = x + 2y + 1$. We can integrate that:

$$\begin{aligned} A(x) &= \int_0^3 (x + 2y + 1)\,dy \\ &= (xy + y^2 + y)\Big|_{y=0}^{y=3} \quad \text{(remember: } x \text{ is constant!)} \\ &= (x \cdot 3 + 3^2 + 3) - (x \cdot 0 + 0^2 + 0) \\ &= 3x + 9 + 3 - 0 \\ &= 3x + 12. \end{aligned}$$

There it is: $A(x) = 3x + 12$. If we check the cases where $x = 0$ and $x = 4$, we get $A(0) = 3(0) + 12 = 12$ and $A(4) = 3(4) + 12 = 24$. (That's what I noted earlier. Did you verify that?)

The volume of the entire solid is now at our fingertips:

$$\begin{aligned} \text{Volume} &= \int_0^4 A(x)\,dx \\ &= \int_0^4 (3x + 12)\,dx \\ &= \left(\frac{3}{2}x^2 + 12x\right)\Big|_0^4 \\ &= \left(\frac{3}{2} \cdot 4^2 + 12 \cdot 4\right) - \left(\frac{3}{2} \cdot 0^2 + 12 \cdot 0\right) \\ &= 24 + 48 = 72. \end{aligned}$$

Wait a minute! That looked like a regular integral. Wasn't it supposed to be a double integral?

Well, it was. Most definitely. We just need to remember where it came from. After all, $A(x)$ was itself an integral. If I put them together, just the way I was doing at the beginning of the chapter, we have the actual double integral we just evaluated:

$$\begin{aligned}\text{Volume} &= \int_0^4 A(x)dx \\ &= \int_0^4 \int_0^3 (x + 2y + 1)dy\, dx.\end{aligned}$$

Yes, we computed a double integral, all right. We just did it in two easy stages. Well, kind of easy.

By the way, mathematicians often like to use a kind of shorthand for double integrals. The notation $\iint_R f(x,y)\,dA$ is often used to represent the double integral of $f(x, y)$ (a function with two variables) over the region R. The dA signifies that it's an integral over an *area*. And, yes, you're right: Mathematicians also compute *triple* integrals and use dV to indicate it's over a *volume*. Very good! (Sorry, we're not doing any triple integrals.)

Cutting corners

We can draw from our example a general approach for finding the volume of a solid object. If the object is based on a region R and bounded above by a two-variable function $z = f(x, y)$, then its volume is given by the double integral

$$\text{Volume} = \iint_R f(x,y)\,dA.$$

If we are given the function $f(x, y)$, then the only problem is working with the region R. It's from R, you see, that we get our limits of integration. For our next example, I'm going to use the triangle in Figure 30-7. It's labeled R because R stands for "region" (not "rectangle") when it's used for a double integral. As you can see, R lies above the interval [0, 2]. That means the x limits of integration have to be $x = 0$ and $x = 2$. The region is bounded below by $y = 0$ (the x axis) and above by the line

Figure 30-7.

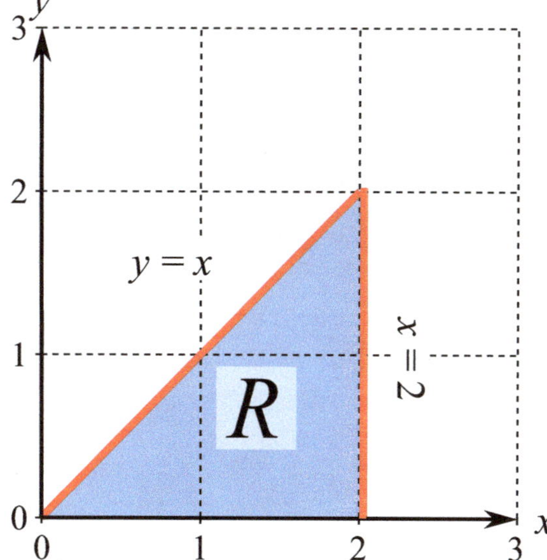

$y = x$, so those will be the limits of integration for y. Now all we need is a function to integrate over the region R.

I propose that we compute the volume of a solid whose base is the triangle R and whose upper surface is a *paraboloid*. A paraboloid is just a parabolic solid of revolution. The equation of the paraboloid I want to use is $z = 9 - x^2 - y^2$. The solid object whose volume we're going to compute is shown in Figure 30-8. Using the two-variable function $9 - x^2 - y^2$ for our integrand and the limits of integration corresponding to the triangle R, we can compute the volume:

$$\begin{aligned}
\text{Volume} &= \iint_R (9 - x^2 - y^2)\,dA \\
&= \int_0^2 \int_0^x (9 - x^2 - y^2)\,dy\,dx \\
&= \int_0^2 \left(9y - x^2y - \frac{1}{3}y^3\right)\Big|_{y=0}^{y=x} dx \\
&= \int_0^2 \left[\left(9x - x^3 - \frac{1}{3}x^3\right) - \left(9\cdot 0 - x^2\cdot 0 - \frac{1}{3}\cdot 0^3\right)\right] dx \\
&= \int_0^2 \left(9x - \frac{4}{3}x^3\right)dx = \left(\frac{9}{2}x^2 - \frac{1}{3}x^4\right)\Big|_0^2 \\
&= \frac{9}{2}\cdot 4 - \frac{1}{3}\cdot 4^3 - \left(\frac{9}{2}\cdot 0^2 - \frac{1}{3}\cdot 0^4\right) \\
&= 18 - \frac{16}{3} = \frac{38}{3}.
\end{aligned}$$

Figure 30-8.

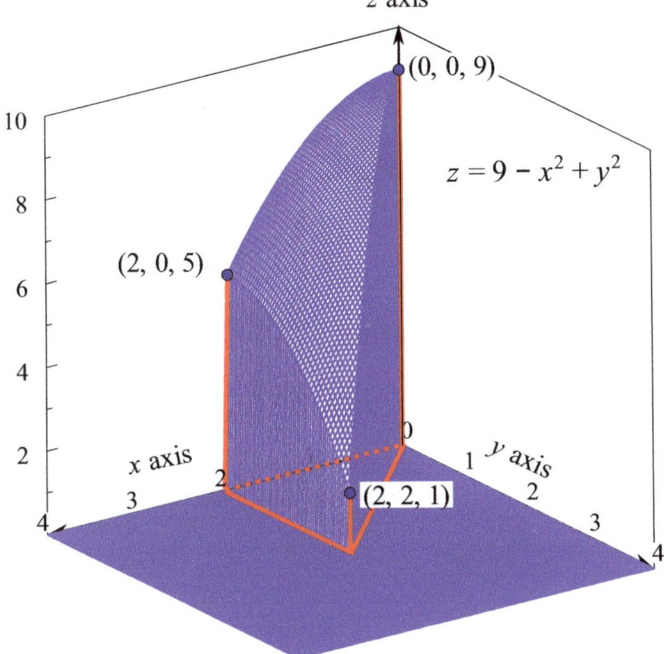

There you have it: A complete calculation of a double integral. Notice that we worked it from the inside out. First, we did the inner integral—the one with the dy—while holding x perfectly constant. Then it was time to do the outer integral—the one with the dx. In a calculus class you would probably learn how to work double integrals in different orders, depending on what's more convenient in a given situation. We, however, are done with this introduction of double integrals. We know enough to move on to the final chapter of our stroll, where we will use double integrals to achieve a point of balance.

31 The Gravity of the Situation

On the average, I mean

We are drawing near the end of our stroll through calculus. In this final chapter, I will lead you down a middling path that avoids extremes. We have, of course, talked about extremes, as when we were working on maximum and minimum values of functions. We will be avoiding those in this chapter because the topic is *averages*. By the time we're done, we'll have met a special kind of average that is, shall we say, the central point of our discussion.

Our first average is the simplest one in calculus. What is the average value of a function $f(x)$ over the interval $[a, b]$? It's defined this way:

$$\text{Average} = \frac{1}{b-a}\int_a^b f(x)dx.$$

A picture will show exactly what we're doing when we average a function this way. The average is simply the height of the rectangle whose area equals the integral of $f(x)$ over the interval. For example, Figure 31-1 shows the result of computing the average of $f(x) = x^2 - 4x + 6$ over the interval $[1, 4]$. I claim that the area under the graph of $y = x^2 - 4x + 6$ from $x = 1$ to $x = 4$ is the same as the area under the graph of $y = 3$ for the same interval. In other words, the average value of $f(x)$ over the interval $[1, 4]$ is 3. Let's do the math.

According to the definition, the average is given by the following on the next page.

Figure 31-1.

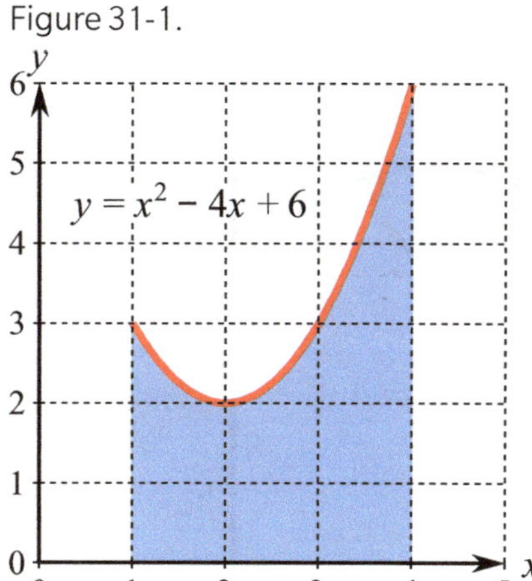

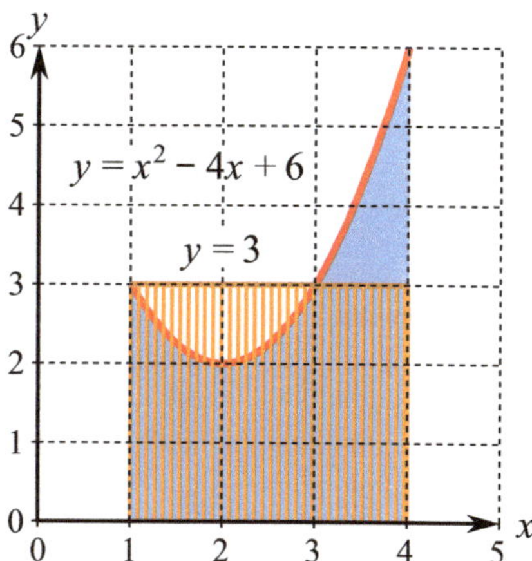

$$\text{Average} = \frac{1}{4-1}\int_1^4 (x^2 - 4x + 6)dx$$
$$= \frac{1}{3}\left(\frac{1}{3}x^3 - 4\cdot\frac{1}{2}x^2 + 6x\right)\Big|_1^4$$
$$= \frac{1}{3}\left[\frac{1}{3}\cdot 4^3 - 2\cdot 4^2 + 6\cdot 4 - \left(\frac{1}{3} - 2 + 6\right)\right]$$
$$= \frac{1}{3}\left[\frac{64}{3} - 32 + 24 - \frac{1}{3} - 4\right]$$
$$= \frac{1}{3}\left[\frac{63}{3} - 12\right] = \frac{1}{3}\cdot 9 = 3.$$

As you can see, this validates my claim that the average of the function is just 3. We can view the average value as the number we would get from measuring the height of the function after leveling out the area into a rectangle whose base is the interval of integration.

Figure 31-2.

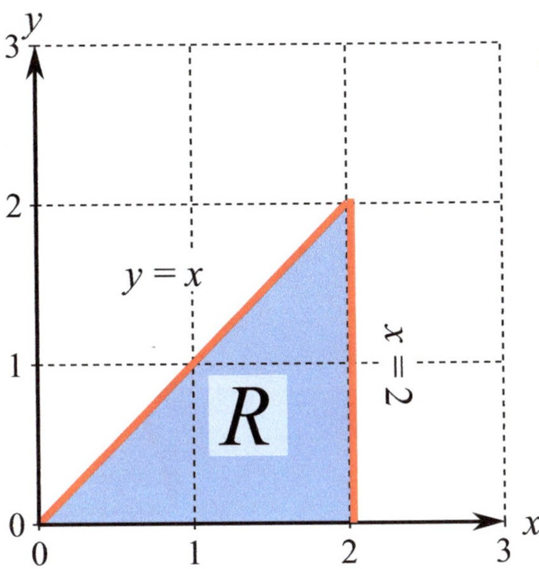

Double down

In the previous chapter, we learned how to compute double integrals. How shall we define the average of a function $f(x, y)$ over a region R?

Let's look again at the volume of the solid region bounded above by the paraboloid $z = 9 - x^2 - y^2$ and based on the triangle R shown in Figure 31-2. We know from our work in the last chapter that its volume turned out to be $\frac{38}{3}$. Since the region R has area 2, how tall would a triangular solid have to be to have the same volume as the paraboloid? We simply compute

$$\frac{38/3}{2} = \frac{19}{3} = 6\tfrac{1}{3}.$$

If you look at Figure 31-3, you'll see that I've drawn in the triangular solid whose height equals the average value of the paraboloid $z = 9 - x^2 - y^2$. If you put together what we just did in a general way, we can see that the average value of $f(x, y)$ over R should be defined as

$$\text{Average} = \frac{1}{\text{Area of } R}\iint_R f(x,y)dA.$$

Figure 31-3.

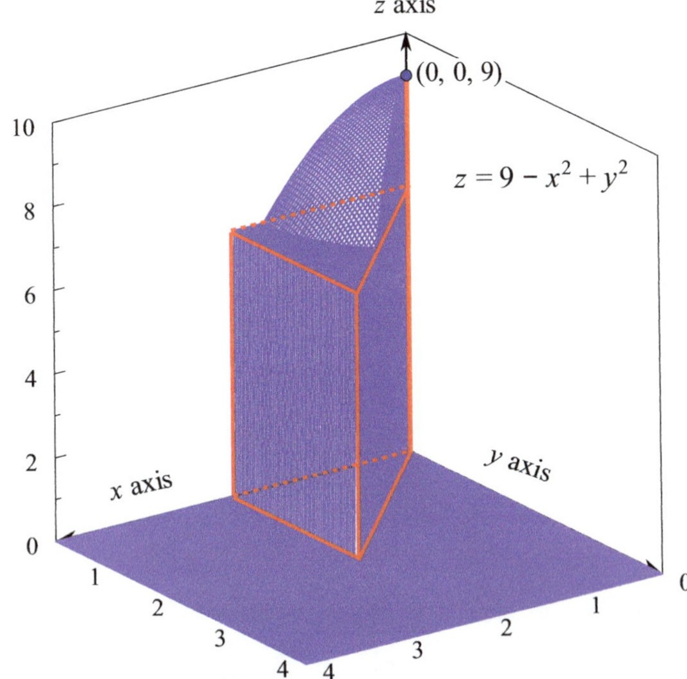

In other words, we compute the volume of the solid, as provided by the double integral, then we divide by the area of the base region R.

We are not, however, bound to regard our computation as given the height of a geometric object. Just as I discussed at the beginning of the book, integrals can be used to represent many different things. For my final examples, I am going to return to the rectangles and triangles of the first two chapters and ask, "What is the average value of their coordinates?" The results will be more significant than you might at first realize.

First up is the 3 by 4 rectangle that I have used so often. There it is as region R in Figure 31-4 again. Each point in R has two coordinates—x and y. I want to find the average x value in the region R and the average y value in the region R. Let's talk about x for a second. The x values range from 0 to 4, of course, so the average x value should be ... what would you say? I think we can agree that we would expect an answer of 2, right? Let's see if our double integral definition of the average of a function makes sense if we apply it to finding the average of x over the rectangle R:

$$\begin{aligned}
\text{Average} &= \frac{1}{\text{Area of } R}\iint_R f(x,y)\,dA \\
&= \frac{1}{12}\iint_R x\,dA \\
&= \frac{1}{12}\int_0^4\int_0^3 x\,dy\,dx \quad (x \text{ is constant for now}) \\
&= \frac{1}{12}\int_0^4 xy\Big|_{y=0}^{y=3}\,dx \\
&= \frac{1}{12}\int_0^4 (x\cdot 3 - x\cdot 0)\,dx \\
&= \frac{1}{12}\cdot\frac{3}{2}x^2\Big|_0^4 \\
&= \frac{1}{8}(4^2 - 0^2) \\
&= \frac{1}{8}\cdot 16 = 2.
\end{aligned}$$

Figure 31-4.

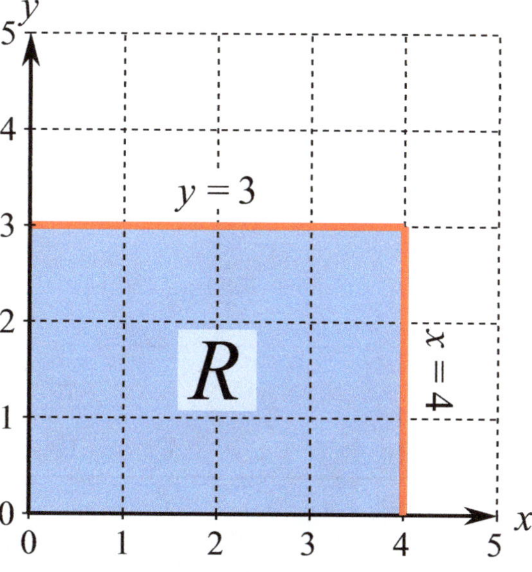

That was quite a bit of work to get a result we never doubted. If we repeat the calculation to find the average of y, we get an equally unsurprising result. Please be patient with me just a moment longer. Our next example will be a bit less obvious. Here's the average value of y for points in the rectangle R:

$$\begin{aligned}
\text{Average} &= \frac{1}{\text{Area of } R}\iint_R y\,dA \\
&= \frac{1}{12}\int_0^4\int_0^3 y\,dy\,dx \\
&= \frac{1}{12}\int_0^4 \frac{1}{2}y^2\Big|_0^3\,dx \\
&= \frac{1}{12}\int_0^4 \frac{1}{2}(3^2-0^2)\,dx \\
&= \frac{1}{12}\int_0^4 \frac{9}{2}\,dx \\
&= \frac{1}{12}\cdot\frac{9}{2}x\Big|_0^4 \\
&= \frac{3}{8}(4-0) = \frac{3}{2}.
\end{aligned}$$

I hope that's what you were expecting for the average value of y. When we put these two results together, we have the point $(2, \frac{3}{2})$. If we graph it with R, as in Figure 31-5, we get the point right in the middle, which is where an average should come out.

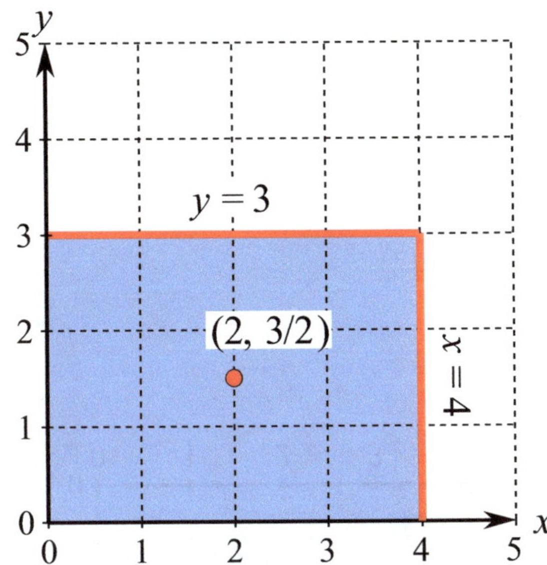

Figure 31-5.

You may be thinking that the rectangle was too easy. Calculus was hardly necessary. I agree. Let's move on to the example that I promised would be more challenging—or at least a little less obvious. This time our region R will be a triangle instead. As x ranges between 0 and 3, y will be bounded below by $y = 0$ and above by $y = x$. Figure 31-6 shows us the region under consideration.

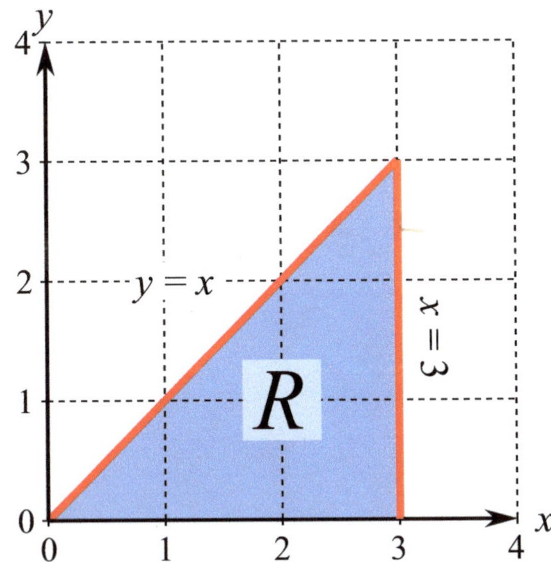

Figure 31-6.

Now I have a question for you: Where would you pencil in the "center" or average point of the triangle? Although x ranges from 0 to 3, we can see that $\frac{3}{2}$ will hardly do as the average value of x. For one thing, there are a lot more points on the right side (where x is larger) than on the left side (where x is closer to 0). The average value of x should be greater than $\frac{3}{2}$. Perhaps you have a good guess. Let's find out for sure with the help of a double integral:

$$\begin{aligned}
\text{Average} &= \frac{1}{\text{Area of } R} \iint_R x \, dA \\
&= \frac{1}{9/2} \int_0^3 \int_0^x x \, dy \, dx \\
&= \frac{2}{9} \int_0^3 xy \Big|_{y=0}^{y=x} dx \\
&= \frac{2}{9} \int_0^3 (x \cdot x - x \cdot 0) dx \\
&= \frac{2}{9} \int_0^3 x^2 dx \\
&= \frac{2}{9} \cdot \frac{1}{3} x^3 \Big|_0^3 \\
&= \frac{2}{27}(3^3 - 0^3) = 2.
\end{aligned}$$

Is that what you expected? Here's the result for y:

$$\begin{aligned}
\text{Average} &= \frac{1}{\text{Area of } R} \iint_R y \, dA \\
&= \frac{1}{9/2} \int_0^3 \int_0^x y \, dy \, dx \\
&= \frac{2}{9} \int_0^3 \frac{1}{2} y^2 \Big|_{y=0}^{y=x} dx \\
&= \frac{2}{9} \int_0^3 \frac{1}{2} (x^2 - 0^2) dx \\
&= \frac{1}{9} \int_0^3 x^2 dx \\
&= \frac{1}{9} \cdot \frac{1}{3} x^3 \Big|_0^3 \\
&= \frac{1}{27}(3^3 - 0^3) = 1.
\end{aligned}$$

When we put the results together, we see that the "center" of the triangle R is (2, 1). I've plotted the point in Figure 31-7 so you can see what it looks like. What do you think?

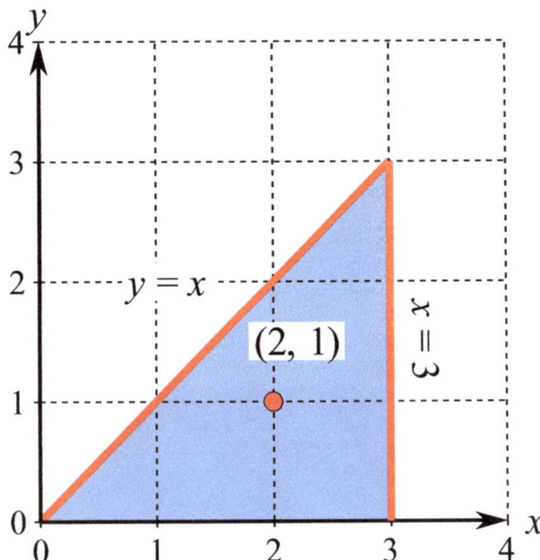

Figure 31-7.

The usual name for the "centers" we just found for the rectangle and the triangle is *centroid* or *center of mass*. Double integrals make it pretty easy to compute a center of mass, but of what earthly good could it be?

Actually, one of the important applications of centers of mass is not earthly at all. It is literally out of this world. When a rocket scientist designs the propulsion system of a space vehicle, it is absolutely crucial that the rocket engine's thrust be applied at the center of mass of the vehicle. If the thrust is off-center, the space vehicle will go into a spin. You certainly don't want to do that accidentally.

The center of mass problem was confronted in two ways by American rocket pioneer Robert Goddard. In his first liquid-fueled rockets, Goddard effectively skirted the issue by putting the rocket motor *above* the full tank. With the heavy tank dangling beneath the rocket (and protected from its exhaust with a conical splash shield; see Figure 31-8), Goddard's device had a natural tendency toward vertical stability. It wasn't as much as Goddard hoped for, but it was a first step.

Figure 31-8.

As his rocket technology grew more sophisticated and the rocket designs began to look more like our modern concept of a space rocket (see Figure 31-9 on the next page), Goddard had to resort to gyroscopic stabilization to keep the thrust of the rocket motor aligned with the center of mass of the vehicle. It's all part of what they call rocket science, and calculus is definitely a part of it.

While the simple regions in our examples are hardly as complicated as space vehicles, they suffice to illustrate the necessity of applying a force at the center of mass. If you were to use Figure 31-7 as a guide and cut a matching triangle out of cardboard or other stiff material, the balancing point of the triangle would be right at the coordinates (2, 1). No doubt it would be difficult to balance the triangle on the point of a pencil or other sharp object, but you wouldn't even have a chance if you tried to do it at any point other than the center of mass. The farther you are from the center of mass, the faster the object will flip and fall.

Figure 31-9.

Parting words

Our stroll through calculus is ending as it began, surrounded by rectangles and triangles. In the beginning, we used the simplest arithmetic to compute some basic quantities and interpret them in the language of calculus. As we continued on our stroll, the tools we used became more sophisticated and we found ourselves able to tackle more complicated problems. Although it was never our purpose to delve too deep, we finished up with the power of double integrals and the importance of centers of mass—or balancing points.

I hope you feel a sense of balance, of equilibrium, at journey's end. We did a little integral calculus, a little differential calculus, and we found that they were opposite sides of a beautifully symmetric structure. Occasionally we were joined by tour guides who shared with us some of their amazing insights, and I hope that I helped you to appreciate their accomplishments.

Our stroll through calculus is over, but what you do next is up to you. Perhaps you are now satisfied that you know a few things about the role that calculus plays in math and science. Perhaps you are still curious and have developed an appetite for even more. I've listed some possibilities for further reading in the afterword and encourage you to explore further.

And with that, your tour guide wishes you farewell and happy wandering.

Afterword

So what's next?

If you want to learn even more about calculus, you have many choices. The classic primer *Calculus Made Easy* by Silvanus P. Thompson was updated by Martin Gardner in a new edition that is readily available. Gardner preserves Thompson's easy-going way of introducing basic calculus topics and adds his own chapters of historical vignettes and mathematical curiosities. There are many thousand-page tomes intended for full-fledged calculus courses in college. I'm naturally rather fond of *Calculus and Analytic Geometry* by Sherman K. Stein and Anthony Barcellos. Our book is out of print now, but cheap copies are available on the used book market.

> *It may be confidently assumed that when this book* Calculus Made Easy *falls into the hands of the professional mathematicians, they will (if not too lazy) rise up as one man, and damn it as being a thoroughly bad book. Of that there can be, from their point of view, no possible manner of doubt whatsoever. It commits several most grievous and deplorable errors.*
>
> Calculus Made Easy
> *Silvanus P. Thompson & Martin Gardner*

If you'd like to become better acquainted with Archimedes, you should read *Archimedes: What Did He Do Besides Cry Eureka?*, written by Sherman Stein. The competition between Newton and Leibniz for the credit of having invented calculus (each one did it independently) is treated in detail in *The Calculus Wars* by Jason Socrates Bardi.

Acknowledgments

The gimlet-eyed mathematician will easily find many infelicities and redundancies in this book, as well as a general lack of rigor. I owe a dedicated band of friends and colleagues for ensuring that all such occurrences were intentional and pedagogically motivated. My former co-author and mentor, Sherman Stein, advised me to maintain a conversational style and to avoid fussing with details that would be more likely to distract readers than enlighten them. My colleagues at American River College, Ted Ridgway, Benjamin Etgen, Brandon Muranaka, and John Burke, indulged me during many conversations and e-mail messages concerning the manuscript. They were invaluable sounding boards and very patient with me. Ted and Benjamin both gave the manuscript a very close reading at various stages and shared detailed lists of comments. Ted even exercised his own substantial writing skills on my behalf and offered alternatives for my clumsier prose passages; I gratefully accepted many of them. John Burke proved that every pair of eyes has its own perspective, especially if they're sharp eyes, as he presented me with a number of typos that had escaped previous readers. My friends Eric Butow, Paul Knox, Kathie Baker, and Tim Feldman are not mathematicians, which made their suggestions all the more significant and useful. One has to admire their patience.

Peter Renz was instrumental in obtaining helpful reviews of the manuscript. Ken Ross graciously agreed to read the manuscript and caught several errors. He also pointed out to me the inescapable logic of Mr. Spock's inevitable encounter with radians.

I thank the Missouri School for the Deaf for its kind permission to use the photograph in Figure 8-10.

The original illustrations were created by the author with CoDraw, a splendid scientific graphing program by Bob Simons of CoHort Software (www.cohort.com).

About the Author

Anthony Barcellos has been on the mathematics faculty at American River College in Sacramento since 1987. He is a graduate of Porterville College, Caltech, Fresno State, and the University of California at Davis. His associate's, bachelor's, and master's degrees are in math and his Ph.D. from UC Davis is in math education. While still a teaching assistant in the UC Davis mathematics department in 1977, Barcellos persuaded the California state legislature to adopt a resolution in honor of the bicentennial of the birth of Carl Friedrich Gauss.

The Mathematical Association of America has honored his expository writing with awards for his *College Mathematics Journal* article on Benoit Mandelbrot's fractal geometry, and his interviews of Mandelbrot, Martin Gardner, and Stanislaw Ulam were collected in *Mathematical People* (Birkhäuser Boston, 1985). Between 1985 and 1990, Barcellos served as editor of *Sacra Blue*, the monthly magazine of the Sacramento PC Users Group, Inc., a non-profit computer organization. The students of American River College chose him as their Instructor of the Year in 1996, and in 2014 the American River College Patrons Club selected Barcellos for its Patrons Chair Award.

In addition to his work in math and math education, Barcellos has been on the legislative staff of the California state senate, served as an analyst for the California state treasurer's office, and reported science stories for the *Albuquerque Journal* on an American Association for the Advancement of Science Mass Media Fellowship. He became a published novelist with the release of *Land of Milk and Money* (Tagus Press, 2012, www.landofmilkandmoney.com), a fictionalized account of his family's experiences as Azorean immigrants in Central California.

Index

A ∫troll through calculus

is not a textbook.

It's an easygoing tour of the main concepts of calculus without fussing over dotting every *i* and crossing every *t*. Despite what many think, the basic content of calculus is understandable at an elementary level. People who aren't afraid of a little high school algebra can discover in these pages why calculus is so important and so powerful—without getting bogged down in theory or subtle computations.

Anthony Barcellos is a math teacher and novelist who has been honored for his teaching by his students and colleagues at American River College in Sacramento and for his expository skills by the Mathematical Association of America. His novel *Land of Milk and Money* (Tagus Press, 2012) was praised as "a full-blooded tale that readers will find insightful, rewarding, and entertaining."

"Tony's ability to work through technical matters and put them down clearly and appealingly is in a class with Martin Gardner's."

—Peter Renz, mathematician and editor

"*A Stroll through Calculus* proves just how simple calculus concepts really can be and that the only thing necessary to understand them easily is a little bit of algebra!"

—Student LF

"The magic behind *A Stroll through Calculus* for me was that its sole intent was to educate a person on the basic foundation of calculus without the assumption of proficiency in advanced algebraic techniques. Recommended for *every* calculus student? Absolutely!"

—Student BT

"The major thing I noticed in *A Stroll through Calculus* was a really great way of being able to communicate to the average reader."

—Student EN

"I wish I had read this earlier, especially during Calculus II, when I was miserably struggling."

—Student OK

"Tony's *Stroll through Calculus* is required reading in my algebra classes. Not only does the short book give my students a sense of the power of calculus, it empowers them to consider a major that requires calculus."

—Professor Benjamin Etgen

www.ingramcontent.com/pod-product-compliance
Lightning Source LLC
Chambersburg PA
CBHW061420210326
41598CB00035B/6280

* 9 7 8 1 6 3 1 8 9 8 0 1 3 *